Evaluation and Assessment for Conservation

Ecological guidelines for determining priorities for nature conservation

IAN F. SPELLERBERG

Senior Lecturer in Biology, and Director of Studies for Environmental Sciences, University of Southampton

Foreword by Jeffrey A. McNeely
Chief Conservation Officer, IUCN

CHAPMAN & HALL
London · Glasgow · New York · Tokyo · Melbourne · Madras

Published by Chapman & Hall, 2-6 Boundary Row, London SE1 8HN, UK

Chapman & Hall, 2-6 Boundary Row, London SE1 8HN, UK

Blackie Academic & Professional, Wester Cleddens Road, Bishopbriggs, Glasgow G64 2NZ, UK

Chapman & Hall GmbH, Pappelallee, 69469 Weinheim, Germany

Chapman & Hall Inc., One Penn Plaza, 41st Floor, New York, NY 10119, USA

Chapman & Hall Japan, Thomson Publishing Japan, Hirakawacho Nemoto Building, 6F, 1-7-11 Hirakawa-cho, Chiyoda-ku, Tokyo 102, Japan

Chapman & Hall Australia, Thomas Nelson Australia, 102 Dodds Street, South Melbourne, Victoria 3205, Australia

Chapman & Hall India, R. Seshadri, 32 Second Main Road, CIT East, Madras 600 035, India

First edition 1992
Reprinted 1994

© 1992 Ian F. Spellerberg/Evaluation & Assessment for Conversation

Typeset in 10/12pt Sabon by Intype, London

Printed in Great Britain at the University Press, Cambridge

ISBN 0 412 44280 9

A catalogue record for this book is available from the British Library

Library of Congress Cataloging-in-Publication Data available

Cover picture. The maps are based on the decline and fragmentation of woodlands in the County of Warwickshire, England (after Thorpe, H in Hawkes, J.J. Conservation and Agriculture, 1978, pp. 17-44, Duckworth, London). See p.12 & p.60.

Contents

Foreword

Most politicians have jumped on the conservation bandwagon, and nobody running for public office these days can afford to take an overtly anti-environment stand. The fascination that children have for nature, the generous donations people make to conservation organizations, the votes cast for 'Green Parties,' the continuing popularity of zoos and wildlife films, and the strong sales of books about the environment all provide evidence to politicians that the general public supports the idea of conservation. Conservation has become a major issue for governments. No longer is it necessary for conservationists to campaign for getting the cause on the agenda: it is already there, at least as a talking point.

The issue now is how to convert this generalized interest into real action. And among the many priorities competing for attention, how is a government (or a private organization) to decide what to do first? From a very limited budget – for budgets will always be limited – what is the package of activities that is most likely to lead to the results that the public wants?

Ian Spellerberg attempts to address these questions which are at the heart of modern conservation action. It is relatively easy to prescribe useful activities that will benefit both the environment and the public at large. But it is far more difficult to devise actions that will meet these two conditions **and** be acceptable to the local people whose behaviour is expected to change. A farmer who has grazed his sheep for decades on a moor will not take kindly to having his pastures converted into a strict nature reserve for the pleasure of urban birdwatchers, and hunters are notoriously steadfast in defence of their sport.

Some high-minded individuals may be willing to forego some personal gain for the general good – the higher prices paid for some 'green products' is an indicator. But many others will need to be convinced in economic terms that conservation provides real benefits. And since those seeking to exploit nature can marshall ample economic justification in support of their position, conservationists must use similarly rigorous approaches to support less abusive uses of the resources if they are to win the day.

Ian Spellerberg provides a rich buffet of such approaches. Far more than a random sample of savoury dishes, he provides a balanced diet, leavening economics with ecology, endangered species with energy-based evaluation, exploitation with amenities, environmental impact assessment with historical perspectives. Such a meal enables the reader to enquire into the implications of declining populations of wildlife for entire ecosystems:

woodlands, riverine habitats, and even urban areas (where, as he points out, a surprising number of hardy species are able to prosper). Helping us digest a veritable feast of intellectual nourishment, Ian Spellerberg provides the basis for making decisions about how to invest in conservation.

One of his main ingredients is economics, since most politicians consider this dismal science to provide their most useful allies. But he does not accept at face value the assertions of mainstream economists, pointing out that the standard economic models fail to give sufficient weight to long-term benefits; that approaches to assessing the economic values of natural processes such as watershed protection or amelioration of climate remain rudimentary at best; and that the aesthetic, ethical, cultural, and scientific considerations that must be part of the economic equation are usually notable by their absence.

Ian Spellerberg is similarly cautious about ecology, his other main ingredient. He recognizes that even in a data-rich environment like the UK, considerable uncertainty still surrounds the task of assigning qualitative and quantitative values to ecosystem functions. Nobody really knows enough about any gene, species, or ecosystem to be able to calculate its ecological worth in the larger scheme of things. Further, the reasons for the existence of species and ecosystems may be far more subtle than simply supporting the desires of the current generation of consumers, so a simple measure of direct utility is not sufficient. Species important to human welfare are not limited to wild relatives of agricultural crops, or species that are harvested for food, fuel, or medicine; 'low status' life forms such as earthworms, bees, and termites may be even more important to society in terms of the role they play in maintaining healthy and productive ecosystems. Continuing on our current path of over-consumption and mindless pollution is all too likely to threaten especially these more covert species whose contribution to ecological processes is not obvious, even to ecologists. Some of these species may be gone before anyone even realizes that they were going, or why they were here. In essence, we are embarking on a one-way voyage to an uncertain and less diverse future, without giving very much consideration to the consequences.

Realizing that uncertainty is inevitable when altering ecosystems, how can society make the best decisions about how to allocate scarce resources to conservation? How can the 'precautionary principle' be put into practice within the constraints of government budgets and competing priorities? Ian Spellerberg helps us to open our eyes, and to understand the many variables that must be considered in making decisions about conservation.

Society decides what is important through the political process, which in turn is informed by what the public knows about the issues to be addressed, the options available, and the likely results of those actions. *Evaluation and Assessment for Conservation* will help ensure that the choices affecting the remaining natural habitats are informed ones, and that nature will greet our choices with the expected exuberance of wildflowers, plentiful clean water, abundant wildlife, and a better quality of life. Should society settle for anything less?

Jeffrey A McNeely
Chief Conservation Officer

Preface

In 1981 my small book *Ecological Evaluation for Conservation* was published. At that time there were many questions being asked about the ecology underlying some of the methods of evaluation. One reason for writing that book was that I believed more research on ecological evaluation could have a greater and better role to play in conservation and in land use planning. In the last ten years I and many others have researched or have supervised research on the ecological basis of evaluation and assessment of species and communities. Consequently there have been many interesting, theoretical and practical developments; so many that it was easy to justify a text which explores ecological evaluation and its applications in depth. Of particular interest to me is that ecological evaluation and assessment has, of necessity, become an interdisciplinary subject.

This book is a personal view of ecological evaluation and of necessity makes use of a limited range of selected examples. It has been written not only to provide a rationale underlying evaluations and assessments but also to explore the applications of this topic. I would hope that I have provided an applied ecology text as well as a handbook on ecological evaluation for practitioners (from a range of backgrounds) involved in evaluations and assessments. However, because many aspects of evaluations and assessments are described and because there are many methods I felt it necessary to conclude the preface with an 'executive summary' which may help the reader to identify the most important applied aspects.

The world is witness to losses in biodiversity far greater than ever before experienced. Those losses are to our detriment. The quality of life of every person on this earth and of future generations is being eroded by these losses in biodiversity. The underlying rationale for assessment and evaluation of species and communities is that it would be impractical to protect and conserve all species and all communities with an equal amount of support and endeavour. There is simply not enough money and scientific expertise to do this. That being the case, there has to be some kind of selection which favours greater protection or greater efforts to conserve and protect certain species and communities. That is what the science of evaluation and assessment is all about; evaluation is concerned with the methods and criteria used to make selections, to draw up priority lists and to identify species and regions which are in need of greatest protection; assessments are undertaken either to identify the nature and extent of impacts on wildlife or to identify species and habitats sensitive to impacts.

The five essential topics described in this book are as follows:

1. The different methods of placing a value on wildlife: notional monetary values (p. 66), economic values (p. 69), amenity values (p. 70), energy-based values (p. 75) and replacement values (p. 76).
2. The assessment methods used in connection with designating species for protection by law and for inclusion in red books (books with lists of the most endangered species). An appraisal of the methods is given on p. 84.
3. The ecological assessment of biotic communities, of which there are many, are described for various taxonomic groups and habitats in Chapter 5. There is an appraisal of the methods on p. 159 and a conceptual plan for ecological evaluations on p. 238.
4. Assessment and selection of nature reserves, landscape reserves and other protected areas is discussed in Chapter 6 and an appraisal of the methods is given on p. 198.
5. The role of ecological evaluation and assessment of biotic communities in planning and in environmental assessments is given in Chapter 7 and there is a operation framework on p. 228.

An overall discussion of the state of the art of evaluation and assessment for conservation has been left to Chapter 8.

A prerequisite for successful and useful evaluations and assessments seems to be at least a basic appreciation of the principles of ecology. Chapter 2 is therefore devoted to a very brief explanation of those aspects of ecology which underly and are essential for the good practice of ecological evaluation and assessments of species and communities.

There are three appendices: a short bibliography (supplementing the often specialist, scientific references at the end of each chapter); a list of the addresses of most organizations mentioned in the text; and a glossary of scientific terms used in the text.

Ian F. Spellerberg
Southampton

Acknowledgements

Many people and many organizations have helped in the writing of this book. I am most grateful for the guidance and encouragement of Dr R.C.J. Carling of Chapman & Hall. I am especially grateful to the following people who have either kindly allowed me to use published material or who have contributed to the text: George A. Antonelis (USA), Bruce Aylward (UK), Leo Batten (UK), Ronald E. Beiswenger (USA), A.E. Brown (UK), Pam Copson (UK), Michael N. DeMers (USA), John Dover (UK), Pete Evans (UK), Rod Everett (UK), Garth Foster (UK), Tony Gent (UK), Andrew W. Gilg (UK), David R. Given (New Zealand), Frank Götmark (Sweden), Alan Gray (UK), Steve Hardes (UK), Lesley Haskins (UK), H.J. Harris (USA), Rodney Helliwell (UK), James M. Herkelrath (USA), Mike Jacobs (UK), Timothy H. Johnson (UK), Rob Lesslie (Australia), Martin Luff (UK), Jeffrey McNeeley (Switzerland), Brian A. Millsap (USA), David A. Norton (New Zealand), Jay O'Keeffe (South Africa), Colin Ogle (New Zealand), Margaret Palmer (UK), George Peterken (UK), Maurice E. Pickering (UK), Bob Pressey (Australia), Rosemary Purdie (Australia), Charlie Rugeroni (UK), Jack Sinden (Australia), George H. Stankey (USA), H. Sukopp (Germany), William Tans (USA), Walter Tips (Thailand), Colin R. Tubbs (UK), Louise Turbin (UK), A. Ulph (UK), R.A. Vagg (UK), Paul Vickerman (UK), Malcolm Vincent (UK), S.D. Ward (UK), Nigel Webb (UK), Gary Windsor (Australia), R. Wittig (Germany).

The Boldrewood Library at the University of Southampton kindly helped many times and I am most grateful to all the staff at that Library. The following organizations have generously provided information for the book: Arboricultural Association (UK), Dorset Trust for Nature Conservation (UK), IUCN Environmental Law Centre (Germany), IUCN Secretariat (Switzerland), Florida Game and Fresh Water Fish Commission (USA), National Oceanic and Atmospheric Administration (US Department of Commerce), Natural Environment Research Council (UK), The Royal Society for the Protection of Birds (UK), The Wildlife Society (USA), World Conservation Monitoring Centre (UK and Switzerland), World Resources Institute (USA), World Wide Fund for Nature (UK), World Wide Fund for Nature International (International Secretariat, Switzerland).

I dedicate this book to my mother, Mary Elizabeth Spellerberg née McLean (1911–1987): her love and appreciation of nature has been my inspiration.

1

The importance and value of biodiversity

1.1 INTRODUCTION

The term wildlife has been used to refer to species of wild plants and animals (but might not necessarily include organisms such as bacteria, viruses, moulds and other life forms which are neither plant nor animal). Nature is a term which infers not only life forms but also the interactions between those life forms. Biodiversity is a comprehensive word for the degree of nature's variety, including both the number and frequency of ecosystems, species and genes in a given assemblage (McNeely, 1988). It is a word which embraces both species richness and genetic diversity, both of which are being threatened throughout the world. Species extinctions and a reduction in genetic variability is taking place at rates never before witnessed. These losses can be attributed to various factors including pollution, physical disturbance, exploitation for food and other uses, deliberate extirpation, and habitat loss and fragmentation; the last of these is undoubtedly the most serious threat. Symptomatic of the rapidly growing human population, pollution, extinctions and habitat loss are all contributing to a diminished quality of life for future generations.

Concern about the impact of the human race on the environment is not new. For example, in 1864, George Marsh was very perceptive about the way humankind was changing the face of the earth and pointed out the dangers of overexploitation of nature. Since that time there has been a slow but expanding concern about the losses in nature and the deteriorating environment. In addition to many books advocating conservation of biodiversity, there has recently been a growth in the number of statements from governments and international organizations about the need to conserve biodiversity. For example, the 1980 World Conservation Strategy, the United Nations 1982 Charter for Nature, the 1987 World Commission on Environment and Development publication *Our Common Future* and the 1989 World Resources Institute publication *Keeping Options Alive*. The year 1991 saw many interesting publications relevant to conservation on a world scale. Those publications included the National Science Foundation's *Loss of Biological Diversity: a global crisis requiring international sol-*

utions, the Ecological Society of America's *Sustainable Biosphere Initiative: an ecological research agenda*, the ODA's *Biological Diversity and Developing Countries, issues and options* and the sequel to the World Conservation Strategy. A document currently in preparation (1991) is the *Global Biodiversity Conservation Strategy*, a collaborative effort from the World Resources Institute (WRI), the World Conservation Union (IUCN) and the United Nations Environment Programme (UNEP).

In 1992, The United Nations held a UN Conference on Environment and Development (UNCED) in Rio de Janeiro, Brazil (20 years after the Stockholm Conference on the Human Environment). Whereas Stockholm was the incentive for three global environmental treaties, and for the establishment of national initiatives worldwide, UNCED had six components:

1. An 'Earth Charter' or declaration of basic principles to ensure the future viability and integrity of the Earth as a home for mankind and other life;
2. Agreement on specific legal measures including a convention on biodiversity which will be developed before the Conference;
3. An agenda for action ('Agenda 21') establishing the agreed work to be done into the 21st century;
4. New and additional financial resources;
5. Transfer of technology;
6. Strengthening of institutional capacities and processes.

Despite these publications, statements, meetings, formation and growth of many conservation organizations (see Appendix B), the expanding range of national and international wildlife legislation, the many protected areas, and management and restoration programmes, we have gone beyond the stage when conservation can balance the losses in nature.

For the last 20 000 years, humans have modified and changed the natural world. There are few areas in the world not affected by people but some areas are still largely unmodified in that they still retain climax communities (see Appendix C for definition of terms). Natural areas are very rare in industrialized countries if present at all. Although some say that we have only cultural landscapes, those landscapes and the artificial and managed ecosystems do support wildlife of varying conservation interest. The conservation of natural and semi-natural areas together with all forms of life is important to all people and future generations. However, limited resources of money and expertise make it impractical to have a fully comprehensive programme of conservation on a worldwide basis. That poses a problem and choices have to be made regarding where best to direct conservation efforts.

That is the subject of this book, how to evaluate and/or assess species, biotic communities and natural and semi-natural areas for conservation. The theme is how to identify where it is best to direct resources for conservation and the main topics are: (1) evaluation of species and evaluation of natural or semi-natural areas for conservation; and (2) assessment of areas for safeguarding and protection from damage.

1.2 LOSSES IN BIODIVERSITY

About 1.6 million species have been named and there has been much speculation about the total number of species. Some estimates of the total number of species on Earth range between 5 and 10 million. However, those estimates are based on 800 000 described insect species and there may be many more unnamed insects and other arthropods (Erwin, 1982; Gaston, 1991). There has been some criticism of Erwin's estimates of number of species but he has made it clear that the figure of 30 million species was suggested as a testable hypothesis (Erwin, 1991). Recent research on the number of species of fungi has added to the debate about numbers of species: there may be as many as 1.6 million species of fungi rather than the 69 000 or so currently recorded (May, 1991). Meanwhile, the rate at which species are becoming extinct, as a result of exploitation, is at a level never before witnessed. Myers (1983) has suggested that an extinction rate of at least one species a day would not be unrealistic, possibly one every hour.

Why is there a growing loss in biodiversity? The cause is the rapidly expanding world population. The symptoms can be grouped under three broad headings:

1. Pollution, disturbance and other human-made perturbations;
2. Excessive exploitation of species and natural areas for food, collections, materials and other purposes (and direct extirpation of pests);
3. Reduction and fragmentation of natural areas (habitats and ecosystems).

1.2.1 Pollution and related effects

Plant and animal species and habitats are increasingly being harmed by industrial activities and pollution. Excessive use of agrochemicals, the incidence of oil spills and the spread of acid rain are just three examples. In 1962 Rachael Carson published her account (*Silent Spring*) of the dramatic effects of pesticides on wildlife and forecast conditions where many of the well-known forms of wildlife would become extinct. Perhaps somewhat emotive in parts, *Silent Spring* nevertheless generated a lot of discussion, but more importantly it served to initiate much needed research and monitoring of the effects of agricultural chemicals. Ratcliffe's classic and fascinating work on the cumulative effects of DDT on birds of prey, undertaken from 1947 onwards, was just one of the areas of research which led to identification of the now well-known egg shell thinning problem (Ratcliffe,

1980). The occurrence of oil spills around the world (from the coasts of Antarctica to the Arctic ocean) has prompted not only large research programmes of research on the effects of oil spills on wildlife but has initiated monitoring programmes to follow the nature and rate of colonization by plants and animals after an oil spill. Acid rain has become a household word and in the late 1980s the effects of acid rain on European forests and lakes were being widely debated and researched. However, the spread of acid rain over Europe was being observed in the mid 1950s and it has taken all this time to identify sources of acid rain and initiate programmes to combat this form of pollution.

One form of pollution affecting wildlife which has been subject to limited research is litter and debris. Figure 1.1 shows that the effects of litter on wildlife is in many forms. Of considerable concern has been the increase in the amount litter or debris at sea (O'Hara *et al.*, 1988). The use of non-biodegradable materials has become so common at sea that it has resulted in significant mortality in bird, fish and marine mammal populations (Wallace, 1985). Concern about the amount of marine pollution has led to international legislation and regulations. In the USA, the response to apparent increases of marine litter has resulted in workshops and research programmes on the fate and impact of marine debris. For example, the US Department of Commerce National Oceanic and Atmospheric Administration has established a Marine Entanglement Research Programme.

1.2.2 Exploitation of biodiversity

Examples of the exploitation and impact on plant and animal populations by the human race are many and varied. The extinction of the dodo as a result of exploitation for food, attempts to reduce populations of so-called vermin such as wolves and coyotes, the hunting of elephants for ivory, the exploitation of large cats for the fur trade, the exploitation of cacti and orchids for botanic gardens and private collections, and the exploitation of whales and fish for food are a few examples. Each year the trade in wildlife affects about 1 million orchids, 40000 primates, 350 million tropical fish, 3 million live birds and hundreds of thousands of reptiles.

Indirect exploitation of plants, animals and micro-organisms has resulted from excessive use of forests. Tropical forests are often thought of as being the richest areas but it should not be forgotten that wetlands and coral reefs are also species-rich (Reid and Miller, 1989). In this brief account, we look at forests and wetlands only. The loss of forests throughout the world has often been illustrated with accounts of losses of the tropical forests, many of which support large numbers of species and also large numbers of centres of endemism. The tremendous species richness of tropical rainforests, for example, was vividly illustrated by Gentry (1988) who recorded 300 species of trees in just one hectare of Amazon forest. The

Figure 1.1 (a) Effects of litter and debris on wildlife. (a) Sea lion (*Zalophus californianus*) with a plastic can holder around its neck (George A. Antonelis)

Figure 1.1 (b) Sea lion killed by a rope (R. Day). Photographs (a) and (b) were kindly made available by James M. Herkelrath of the NMFS Marine Entanglement Research Programme.

Figure 1.1 (c) Shrew (*Sorex araneus*) trapped in a discarded yogurt carton

Figure 1.1 (d) Sand lizard (*Lacerta agilis*) trapped in a discarded bottle

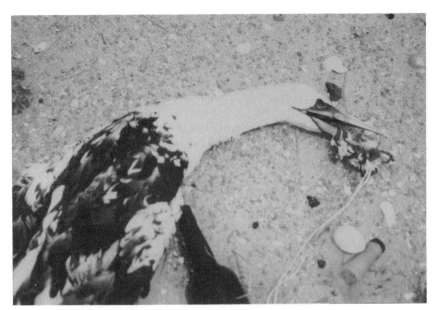

Figure 1.1 (e) Gannet (*Sula bassana*) entangled in line

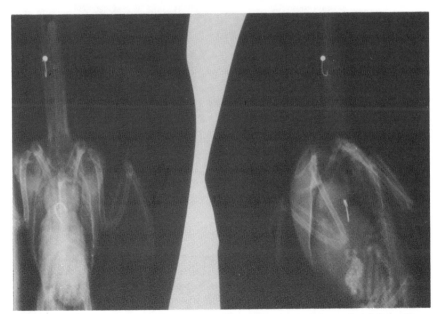

Figure 1.1. (f) Moorhen (*Gallinula chlorpus*) with ingested discarded fishing line.

losses in biodiversity in Central America have been much publicized but it is nevertheless the advances in remote sensing which have provided us with techniques to show the scale of forest clearance (Figure 1.2). The visual evidence is striking but what about quantifying these losses and more importantly are there reliable estimates of the rate and extent of deforestation? This question has been addressed by Fearnside (1990) in a valuable and detailed appraisal of the available estimates. Advanced remote sensing techniques, especially by way of satellites have provided much of the raw data for reliable estimates. Although it is not wise to quote out of context (especially in this case where a subject has been treated in depth) Fearnside

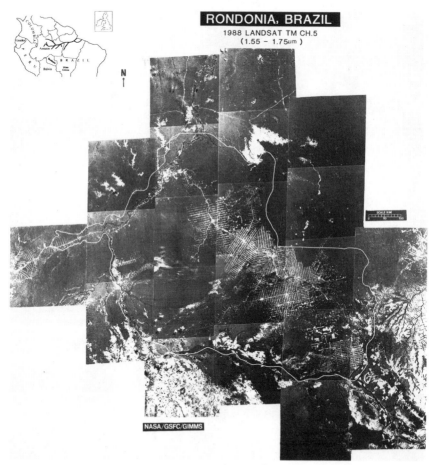

Figure 1.2 Part of Rondonia, Brazil from a false colour composite of the LANDSAT MSS study area described by Nelson and Holben (1986). The linear features are colonization clearings: the midlines of the linear features are 4 km apart. Photograph Kindly provided by Brent Holben. The insert map shows the location of Rondonia and also the size in relation to Britain.

Table 1.1 Loss of rainforests does not occur only in the Amazon or other tropical areas. Rainforests are found in temperate as well as tropical regions and the exploitation there is equally serious as in other parts of the World. Modified from Lewis (1990) after Myers (1989)

---

1. AUSTRALIA

Australia is the only industrialized country to possess major tracts of tropical rainforest. It also possesses areas of endangered cool temperate rainforests (in Tasmania) and rare wet sclerophyll (eucalypt) forests on the east coast. It is expected that very little old-growth forest will survive outside of protected areas by the year 2010.

New South Wales
Forest type: dry and wet sclerophyll forest (more than 150 species of forest eucalypt trees).
Area covered: 630 000 ha.
National parks etc.: 63 000 ha.
Area cut each year: 5450 ha (28.5 football pitches a day)
Method: clear-felling; woodchipping.
Species threatened/dependent on forest ecosystem: yellow-bellied glider (*Petaurus australis*), feather-tailed glider (*Acrobates pygmaeus*), greater glider (*Schoinabates volans*), tiger quoll (*Dasyurus maculatus*), southern brown bandicoot (*Isodon obesulus*), little red flying fox (*Pteropus scapulatus*), powerful owl (*Ninox strenua*), gang gang cockatoo (*Callocephalon fimbriatum*).

Tasmania
Forest type: cool temperate rainforest.
Area threatened: 0.4 million ha.
National parks etc.: 1.4 million ha.
Area cut each year: 20 000 ha (100 football pitches a day).
Method: clear-felling, woodchipping (of eucalypt and mixed forest).
Species threatened include: grey goshawk (*Accipter novae-hollandiae*).
Species dependent on forest habitat: eastern pygmy possum (*Cercartetus nanus*), sugar glider (*Petaurus breviceps*), yellow-tailed black cockatoo (*Calyptorhynchus funereus*), wedge-tailed eagle (*Aquila audax*).

2. CANADA

The world's second largest timber exporter after the US. Still almost 50% forested, though most old-growth forest outside protected areas will be cut within the next 20 years. Main area of exploitation is British Columbia.

British Columbia
Forest type: climax temperate rainforest receiving up to 5 m precipitation a year.
Area threatened: 23 million ha (includes all types of forest).
National parks etc.: 186 000 ha.
Area cut each year: 260 000 ha (1353 football pitches a day).
Method: clear-cutting and burning.
Species threatened: Sitka black-tailed deer (*Odocoileus hemionus*).

Table 1.1 continued

Species dependent on forest habitat: bald eagle (*Haliaetus leucocephalus*), caribou (*Rangifer* sp.), grizzly bear (*Ursus horribilis*).

3. **CHILE** (Bio-Bio, Maule regions)

The coastal and Andean rainforests have been decimated, with less than 20% now remaining. In Bio-Bio, 31% of the old-growth coastal forest has been lost in the past 20 years. At present rates, all coastal temperate rainforest in Chile will have disappeared within 30 years.

4. **CHINA**

Now only 9% forested and, in some areas, has lost more than 70% of its original forests. Yet it still possesses areas of subtropical seasonal monsoon forests in the south, mainly in Yunnan, Guanggxi and Guangdong provinces.

5. **NEW ZEALAND** (Nelson, West Coast, Marlborough, south-eastern Otago)

Indigenous forests, comprising dry and moist temperate forests of native beech, kamahi and rata, cover 23% of the country. The area of forest cut each year is 1250 ha (7 football pitches) a day. The forests are exported as woodchips to Japan, where they are turned into paper. An estimated 1350 forest-dwelling birds are killed for every woodchip shipment to Japan, including parakeets (*Cyanoramphus* sp.), robins (*Petroica australis*), riflemans (*Acanthisitta chloris*) and yellow-heads (*Mohoua ochrocepala*).

6. **USA**

Only 5% of virgin forest remains. The US is the world's largest exporter of timber, principally to Japan. Old-growth logging in the Pacific North-west has been characterized as 'the last great American buffalo-hunt'; 87% of the Pacific North-west has already been logged. At present rates, there will be no significant stands of old-growth timber remaining in this region within 10 years outside of protected areas. The Tongass is the last largely intact rainforest in the US, home to the greatest concentration of bald eagles and grizzly bears in the country. It is estimated that 50% of the low-elevation, high-productivity forest area has been lost since 1950.

Alaska: Tongass

Forest type: coastal temperate rainforest, receiving more than 5 m of precipitation a year.

Area threatened: 1.24 million ha.

Area cut each year: 8000 ha (40 football pitches a day).

Method: clear-cutting; 300 miles of new roads built each year.

Species threatened: bald eagle (*Haliaetus leucocephalus*), grizzly bear (*Ursus horribilis*), Sitka black-tailed deer (*Odocoileus hemionus*).

Species dependent on forest habitat: black bear (*Ursus americanus*), mountain goat (*Oreamnos americanus*), moose (*Alces alces*), marten (*Martes americana*), mink (*Mustela vison*), otter (*Lutra canadensis*), Owls, Canada goose (*Branta canadensis*), great blue heron (*Ardea herodias*), and various salmon species.

Ancestral homeland to: Haida and Tingit Indians.

Table 1.1 continued

Pacific North-west US; North California, Washington and Oregon
Forest type: climax temperate rainforest receiving up to 4 m precipitation a year.
Area threatened: 0.8–1.2 million ha (original cover 7.5 million ha).
National parks etc.: 400 000 ha.
Area cut each year: 24 800 ha (129 football pitches a day).
Method: clear-cutting, slash-and-burn, herbicide spraying.
Exported to: Japan (more than 50%), Korea, Taiwan and China.
Species threatened: pileated woodpecker (*Dryocopus pileatus*), northern flying squirrel (*Glaucomys volans*), and species of guillemots.
Species dependent on forest habitat: 200 species of birds, mammals, reptiles and amphibians live predominantly in old-growth forest.

Hawaii: Wao Kele O Puna
There are 10 800 hectares of tropical lowland rainforest in Hawaii. Most of this forest, the last tropical rainforest in the US, is still unstudied, and it serves an important cultural and social function for native Hawaiians. The whole area is threatened by logging and by chipping of forest trees to produce 'alternative' electricity for Hawaii and Japan. The main threat to the forest now is the construction of an initial series of 20 geothermal wells.

---

concludes that 'Forest loss in 1988 was proceeding at 20×10^3 km^2/year; inclusion of estimated cerrado-loss (Central Brazilian scrub savanna) raises the total to 39×10^6 km^2/year, an area almost the size of the Netherlands'.

Growing widespread concern for the conservation of remaining areas of Amazonian forests has led to a workshop to try to determine some priorities for conservation. That workshop produced a map indicating areas of various levels of priority for conservation; about 55% of Amazonia is a priority for conservation. Although the Peruvian Government is already acting on the information provided by this map and although it is hopeful that other governments may also follow (Kuliopulos, 1990), there still remains the fact that there is little guidance for those making decisions about the conservation of the area. The need for effective political decisions, coordination of policy and reviews of management effectiveness seem to have been overlooked. Such aspects of conservation of protected areas are commonly forgotten but have been usefully addressed in Lucas's book (1992) *Protected Landscapes: A guide for policy makers and planners*.

Tropical forests once covered about an estimated 25 million km^2 of the land surface of the earth but by the mid 1970s this had been reduced to about 9 million km^2. Since 1950 Latin America has lost about 37% of its forests, Central America about 66%, Central Africa about 52% and Southeast Asia about 38% (these figures with appropriate references are given in WCED, 1987). However, losses in forests are not restricted to tropical regions and there have been very serious losses of temperate forests in countries such as Canada, Australia and New Zealand (Table 1.1). Also, the losses of threatened plant species in temperate regions is perhaps higher than might be expected: in North America about 10% of the plant species were threatened in 1986 and in Europe the figure was 17% (Cronk, 1988). Although losses of biodiversity in the tropics is a serious problem, we must not forget that the same process is occurring in temperate regions.

The implications of losses in forests throughout the world are many. As well as losses in species (many of which are unknown to science) there are alterations to the local climate and rivers and watersheds are damaged and consequently there is increased erosion. Furthermore, deforestation (the burning and the loss of plant life) causes increases in levels of 'greenhouse' gases such as CO_2. According to Myers (1989), over the period 1979–1989 there was a 41% increase in the release of carbon as a result of deforestation. Disruption of moisture cycling through partial deforestation may result in the steady desiccation of some remaining forests. It is somewhat ironic that recent studies show that financial benefits generated by sustainable use of forest resources tend to exceed those that result from forest conversion (Peters *et al.*, 1989).

The other community to be discussed briefly here are wetlands. Wetlands is a collective term for a range of inland, coastal and marine habitats characterized by permanent water-logging (but see definitions in Appendix C). There are saltwater, freshwater and artificial wetlands and examples include estuaries, coasts, marshes, swamps, lakes, peatland, bogs, fens and artificial reservoirs. Wetlands, especially mangrove swamps, are areas of very high biological productivity and support high levels of species richness. In addition to their biological value, wetlands have many other values and functions which can be evaluated using other criteria (see for example Winpenny, 1991). They are important locations for fisheries, agriculture, water supplies, forestry, recreation and tourism; some also provide flood control and storm protection. Losses in wetlands have been huge; for example, in the USA alone, about 54% of wetlands have been lost, 87% to agricultural development.

1.2.3 Reduction and fragmentation of habitats

The greatest threat to wildlife today is the ever-increasing exploitation of land. In a sense, the niche of humankind is expanding at the expense of most other species. This ever-increasing exploitation is resulting in damage, loss and fragmentation of natural areas in all continuously inhabited parts of the World; it is happening in desert regions, temperate regions and tropical regions. The term insularization sums up the whole process of reduction, fragmentation and isolation. One estimate puts 67% of all endangered, vulnerable and rare species of vertebrates (including fish) as being threatened by habitat degradation, loss and fragmentation (Prescott-Allen and Prescott-Allen, 1978). The reduction in area and fragmentation of biotic communities results in immediate extinctions at the edge of the newly created fragment as a result of physical disturbance and changes in physical features and the microclimate. Dramatic examples of extinctions caused by fragmentation have come from the Minimum Critical Size Ecosystem Project in the Amazon Basin (Lovejoy, 1984; Lovejoy *et al.*, 1986). Lovejoy and his colleagues have reported increased tree mortality and diminishing populations of many species in the smaller fragments as well as total extinction of some plant and animal species. However, there were some interesting exceptions, for example, species richness of butterflies at first declined in forest fragments as the surrounding forest was cleared but then increased to levels higher than that for the undisturbed forest. It appears that whereas forest edge and glade species are favoured by the new conditions in the forest fragments, the forest butterfly communities could be adversely affected (Lovejoy, 1989).

Fragmented and isolated populations become more vulnerable to disturbance and inbreeding can occur in small isolated populations, sometimes leading to loss of genetic diversity (Lande, 1988). Dodd (1990) found that in the case of musk turtles (*Sternotherus depressus*), habitat fragmentation led to population decline, abnormal population structure and made the isolated populations vulnerable to human disturbance. The effects of fragmentation and isolation may also apply to birds and other taxa with seemingly good powers of dispersal. In isolated canyons near San Diego, for example, chaparral birds seem more prone to extinction in the fragmented landscapes because of poor dispersal ability and because the distribution of native predators may also influence species numbers (Soule *et al.*, 1988). In general, insularization can result in the following:

1. Reductions in populations and extinctions of species at the newly formed edge;
2. Losses in keystone species (Paine, 1969) or key species (Lewis, 1978), that is, species on which the ecology of several other species may depend (see glossary in Appendix C);
3. Isolation of some taxa followed by faunal collapse and genetic drift (Schonewald-Cox *et al.*, 1983);
4. Inbreeding (Lovejoy, 1980);
5. Increasing population size of invasive species;
6. Changes in species composition and trophic structure;
7. Increased vulnerability to physical disturbance and pollution.

1.3 BENEFITS AND FUNCTIONS OF BIODIVERSITY

1.3.1 The range of benefits and values

Does nature have a value and is that value dependent on what anyone is prepared to pay as if nature were an item of art (Figure 1.3)? Is the

Figure 1.3 Some wildlife paintings such as Van Gogh's *Sunflowers* seem to be valued very highly, for example this painting sold for £24750000. Is nature like art and worth whatever someone is prepared to pay for it? Photograph kindly provided by Christie's, London.

economic value of nature and the environment best based on what people are prepared to pay for it? To answer these questions it would seem useful to look first at the benefits and functions of nature. In general, four categories of values and benefits of biodiversity can be identified: the ethical and moral values; enjoyment and aesthetic values; material values; and environmental values (Table 1.2). The first two categories are part of the main theme of this book and the last category is discussed in part in Chapter 3. The utilitarian benefits of nature are discussed in a little more detail here.

1.3.2 Material benefits

(a) Material benefits: food

Despite the many material values and functions of biodiversity, concern about losses in biodiversity may not be forthcoming without further elaboration. That elaboration can easily be provided when considering the role of genetic diversity in improving crops and providing natural sources for medicinal drugs.

Table 1.2 Suggested benefits and functions of biodiversity. From Spellerberg and Hardes (1992)

A. **Ethical and moral values**
   1. Intrinsic value of nature
   2. Natural world has value as a human heritage

B. **Enjoyment and aesthetic values**
   1. Leisure activities ranging from bird watching to walking
   2. Sporting activities ranging from orienteering to diving
   3. Aesthetic value by way of seeing, hearing or touching wildlife
   4. Enjoyment of nature depicted in art

C. **Use as a resource for food, materials, research inspiration and education (utilitarian)**
   1. As a genetic resource for some of the following
   2. As a source of food
   3. As a source of organisms for biological control
   4. As a source of pharmaceutical products
   5. As a source of materials for buildings
   6. As a source of materials for making goods
   7. As a source of fuel for energy
   8. Source of working animals
   9. For scientific research
   10. Educational value
   11. Inspiration for technological development

D. **Maintenance of the environment (ecosystems and climates)**
   1. Role in maintaining CO_2-O_2 balance
   2. Role in maintaining water cycles and maintaining water catchments
   3. Role in absorbing waste materials
   4. Role in determining the nature of world climates, regional climates and microclimates
   5. Indicators of environmental change
   6. Protection from harmful weather conditions: wind breaks, flood barriers

---

Much attention has focused on the number of plant species which form the basis for most of the world's food. About 12 000 plant species have been used for food but only about 150 species have been cultivated. Staple diets throughout the world are based on only a few plant species. Some previous estimates suggest that about 90% of the world's food comes from only 20 species and that 75% of human nutrition is provided by just seven species: wheat, rice, maize, potato, barley, sweet potato and cassava. In the most comprehensive research yet undertaken, Prescott-Allen and Prescott-Allen (1991) found that 103 plant species feed the world. The difference between 103 and figures of between 7 and 20 can be attributed largely to methods of calculation: the use of supply rather than production data.

Whatever the actual number of plant species, there is no doubt that the plants which 'feed the world' are represented by many cultivated forms. Cultivation of drought-resistant forms, disease-resistant forms and varieties with high yields requires a diversity of genetic material and that diversity comes from the wild species. Losses of species diminishes the ability to develop new varieties. Cocoa cultivation, for example, suffers from a lack of genetic variation because commercial crops have been developed from only a few wild forms. Unfortunately, large parts of the centre of genetic diversity of cocoa in Columbia, Eucador and Peru have been damaged by explorations for petroleum and exploitation of forests for agriculture.

There may be tens of thousands of plant species that could provide a basis for new innovations in food production (Plotkin, 1988) but regretfully little is known about many potentially valuable plants. For example, there is more to beans than just beans. Common beans, french beans, lima beans and many other beans are found among 30 species of the genus *Phaseolus* which in turn is just one of 32 tribes in one of the subfamilies of the family Leguminosae (Smartt, 1990). The 30 species of *Phaseolus* represent a rich source of genetic diversity which has not yet been fully exploited. More research on such groups as these could lead to discoveries which may have important implications for improved varieties of crop plants.

There are many other examples ranging from the discovery of salt-tolerant species of tomatoes on the Galapagos islands (providing a potential for growing tomatoes in arid regions where brackish water is used for irrigation) to some potentially valuable Amaranth (*Amaranthus*) species, said to be a delicious grain domesticated in pre-Columbian Mesamerica and the Andes. There has been some support for the suggestion that an Amaranth from the Andes (such as achita or quihuicha) could form the basis for a world staple food crop.

(b) Utilitarian benefits: biological improvement and biological control

Despite the large amount of work on pests and diseases of crop plants, there are many pests (some quite new) that have not been studied in detail. One classic example is the brown rice leafhopper (*Niloparvata lugens*) which in the late 1970s became a serious pest of rice crops throughout the more temperate parts of Asia. Lack of knowledge (ecological assessments) about this species of leafhopper greatly delayed the efforts to control the pest. Biological research on pests is as important as the research on species which might be used in controlling those pests. The potential of the world's animals for combatting and controlling pests and disease has hardly been used at all but there are many good examples based on species as well as ecosystems. The vineyards of western Europe, for example were saved from destruction by an insect pest when it was found that the native North American grape vine possessed natural resistance to attack. German and

French vines are now grafted on North American root stocks, thus taking advantage of the qualities of different genetic stock.

Similarly, maize crops in the United States have benefited from two sources of wild material: fungus-resistant genetic material in Mexican stocks, primitive wild, perennial species (*Zea diploperennis*) known only from a few small areas of montane forest in southern Mexico (Tatum, 1971; Iltis *et al.*, 1979).

As well as developing genetic variation within a species, other forms of variation may improve yields. Agricultural landscapes are characterized by monocultures, which in turn provide abundant food supplies for pest species. Use of broad-spectrum insecticides has affected populations of the pest's predators as has also the losses in hedgerows and other habitats in agricultural ecosystems (agro-ecosystems). Increasing the diversity of agricultural landscapes has been found not only to support increased species richness but also to support predators of some of the pest species and beneficial insects including pollinators of the crops. For example, in California wild brambles provide a habitat for wasps which prey on some pests of grapes.

(c) Utilitarian benefits: medical

Although scientists have investigated only 1% of the world's plant species and a smaller fraction of all animal species (WCED, 1987) there are many medical drugs derived from plants (Table 1.3) and that list is growing (Myers, 1979; Wilson, 1988; Reid and Miller, 1989; Schultes and Raffauf, 1990). About 40% of all prescriptions dispensed in industrialized countries have their origins in wild organisms (Farnsworth and Soejarto, 1985) and although there are about 120 plant-derived drugs in use throughout much of the world, these have been obtained from less than 90 species of plants. Only about 5000 of the 250 000–300 000 flowering plants in the world have been studied for their potential pharmacological properties (Abelson, 1990). Similarly only a very small number of marine organisms have been investigated for possible drugs of importance to medicine.

(d) Utilitarian benefits: industrial uses

Biodiversity has made a major contribution to industry. Some 20 000 natural compounds have been identified from nature including common items such as starches, inks, cosmetics, rubber, timber, beverages, lubricants and polishes. Despite the introduction of synthetic materials, many of these items such as rubber continue to depend on supplies from living organisms.

Other benefits, in addition to those described above, are suggested in Table 1.2. Limits of space prevent them from being discussed but it would seem out of place not to mention research and education. Nature provides us with a cornucopia of material for inspiration in education and inno-

Table 1.3 Some of the major drugs which are derived from plants. Reproduced with kind permission of WWF International from the WWF publication *The Importance of Biological Diversity* (P.S. Wachtel)

Drug	Source		Use
Atropine	*Atropa belladonna*	(Belladonna)	Anticholinergic
Caffeine	*Camellia sinensis*	(Tea)	CNS stimulant
Camphor	*Cinnamomium camphora*	(Camphor tree)	Rubefacient
Cocaine	*Erythroxylum coca*	(Coca)	Local anaesthetic
Codeine	*Papaver somniferum*	(Opium poppy)	Analgesic; antitussive
Colchicine	*Colchicum autumnale*	(Autumn crocus)	Antitumour agent
Digitoxin	*Digitalis purpurea*	(Common foxglove)	Cardiotonic
L-Dopa	*Mucuna deeringiana*	(Velvet bean)	Antiparkinsonian
Menthol	*Mentha* spp.	(Mint etc.)	Rubefacient
Morphine	*Papaver somniferum*	(Opium poppy)	Analgesic
Quinine	*Cinchona ledgeriana*	(Yellow cinchona)	Antimalarial; antipyretic
Reserpine	*Rauvolfia serpentina*	(Indian snakeroot)	Antihypertensive
Scopolamine	*Datura metel*	(Recured thornapple)	Sedative
Strychnine	*Strychnos nux-vomica*	(Nux vomica)	CNS stimulant
Thymol	*Thymus vulgaris*	(Common thyme)	Antifungal

vation in research at different levels from the whole organism to molecular biology. Two more recent research areas which have featured prominently in Conservation Biology are DNA or genetic fingerprinting and genetic engineering. The former provides a valuable way of identifying individuals or varieties of species. For example, in Scotland genetic fingerprinting techniques have been used by foresters to select and identify various types of plantation trees. In the USA, genetic fingerprinting of the rare condor has been undertaken so that suitable breeding pairs can be established. Genetic engineering has the potential for further, exciting advances in biological control and increased crop yield but there is concern about the possible effects of releasing genetically manipulated organisms into the environment.

1.3.3 Economic benefits of biodiversity

The resources of nature are finite and for a long time many organizations have advocated the use of nature in a sustainable manner (that is the use of plants, animals and ecosystems in such a way that the quality of life of future generations is not impaired). Several publications, particularly *The World Conservation Strategy* (IUCN, 1980) and the *Our Common Future* (WCED, 1987) have highlighted the relevance of sustainable development to the world's economy. This stance was echoed in the UN World Charter for Nature proclaimed in 1982 and has also been taken up by the World Bank (1988), The Worldwatch Institute (*State of the World Reports*) and the International Institute for Environment and Development (*World Resources Reports*). The role of ecological science in wise management of resources was justly advocated by Lubchenco *et al.* (1991) who also emphasized the need for an interdisciplinary approach (interfacing between ecological research, education and environmental decision making) as a basis for sustainable biosphere initiatives.

This concern for sustainable development has focused largely on biodiversity which has direct economic benefits but recently the indirect values of biological resources have been the focus of attention. The direct values can be divided into consumptive, productive and non-consumptive values (Table 1.4). In that example of typology of biodiversity values there is the non-use value, that is intrinsic values attached by individuals to the continued existence of certain species such as elephants and turtles.

Table 1.4 An economic approach to types of biodiversity values and an application to wetlands

Value type	Subtype	Example
Use values		
A. Direct	Consumptive	Home consumed forest fruits
	Productive	Plant breeding
	Non-consumptive	Tourism
B. Indirect		Ecological process
C. Option values		Potential value of medicinal drugs
Non-use values		Existence value of certain species

Application for wetlands (from Aylward, 1991):
Use values
A. Direct: fuelwood, fish, wheat, rice, soyabeans, cowpeas, shrubs, grasses, bird viewing, water transport etc.
B. Indirect: groundwater recharge/discharge, flood and flow control, shoreline/bank stabilization, sediment retention, nutrient retention etc.
C. Option values: environmental functions threatened with conversion etc.
Non-use values: Birds, wetland ecosystems etc.

Biodiversity which is of both direct and indirect value are resources, commodities that have a direct or indirect monetary value to man. If that approach is accepted, then there must be examples of non-resources but it does not necessarily follow that non-resources have no value. That is, we can imagine some kind of biodiversity which is a non-resource, a resource that has no demonstrated or conjectural value to man (Ehrenfeld, 1976, 1981). The broad economic values of non-resources (Table 1.5) as identified in Ehrenfeld's review of economic values of non-resources were used as a basis to show that conservation cannot rely solely on economic and ecological justifications.

Table 1.5 Suggusted economic values of non-resources. (From Ehrenfeld (1981) with permission of the author)

1. Recreational and aesthetic values:
   Includes scenic views, hiking, camping, sport hunting
   Tends to favour large conspicuous mammals
2. Undiscovered or underdeveloped values:
   With only 1% of plants having been studied, the potential value of plants is immense
3. Ecosystem stabilization values:
   Intensively managed monocultures have high yields but at the risk of disease and attack by pests
   Increased species richness could help to reduce pest and disease infestation
4. Examples of survival:
   It may make good sense to look at successful natural communities for clues about organization of traits leading to persistence or survival
5. Environmental baseline and monitoring values:
   Monitoring ecological change requires baseline data and is particularly appropriate to conservation.
6. Scientific research values:
   For example, *Xenopus* (the clawed toad) has been of considerable use to embryological research
7. Teaching values:
   Natural areas such as woodlands and unimproved meadows for ecology teaching
8. Habitat reconstruction values:
   Examples of reconstructed ecosystems are rare and the reconstruction requires a model
9. Conservation value:
   Avoidance of irreversible change
   Human-made, irreversible change of the loss of biodiversity may result in as yet unknown risks or damage to the quality of human life

1.4 EXPLOITATION: A WAY TO CONSERVE WILDLIFE AND NATURAL AREAS?

There is a view that rather than protecting wildlife from any form of exploitation it is better to allow sustainable exploitation because that will then provide a basis for conservation; that is, unexploited wildlife is a non-resource and therefore valueless. This was a theme which emerged from one of the workshops held at the 18th General Assembly of the IUCN (Perth, 1990). Two advocates of the widely held belief that wildlife should pay its way are Graham Child and Brian Child, who believe that our failure to conserve wildlife and natural areas has resulted from 'centrally regulated management of game'. That approach relies on 'protectionist devices and policing' – an approach that is flawed because it fails to provide the incentives which are essential for preventing changes in land use such as creating agricultural land from natural and semi-natural areas.

This kind of economic approach to wildlife management is apparently becoming increasingly popular among environmental economists (Lavigne, 1991) but is not a new idea and indeed has had support from conservationists for more than 40 years (e.g. see Leopold, 1933). On the surface, convincing arguments can be put forward to support the idea that economic components of biodiversity can function without the uneconomic components but on closer examination those arguments do not take into consideration the interdependence and interrelationships of species (both economic and uneconomic) nor do they consider potential benefits or unknown values. For example, Lavigne (1991) reported that, the Childs claim that 'the loss of biological diversity that has accompanied agriculture and other economic developments has often seemed beneficial and has been sustainable for hundreds of years'. That may be true for small-scale agriculture but it is not true for large-scale and intensive agriculture which has become so reliant on chemicals and which results in land being lost to agriculture because of the deteriorating quality of the soil and because of damage to watersheds.

1.5 CONSERVATION IN ACTION

1.5.1 *Ex situ* or *in situ* conservation?

It can not be denied that zoological and botanical gardens have made an important contribution to conservation (ex situ conservation). Botanic gardens, for example, act as a midway house in making genetic resources available for the use of people. Some captive breeding programmes have been very successful but they do depend on carefully planned and well-maintained communication networks and databases. Notable in this respect is the International Species Information System (ISIS), a central computerized database designed by Ulysses Seal in the USA, which has its base at the Minnesota Zoological Gardens, USA. The use of ISIS has greatly improved data storage and communication and ISIS has also led to the development of ARKS (Animals Records Keeping System) and the studbook programme SPARKS (Single Population Analysis and Records Keeping System), but how are species best selected for captive breeding? Seal (1985, 1988) has argued that species can be selected for captive breeding by way of a programme called Population Viability Analysis (PVA). In brief, information about the demography, population dynamics and ecology of a species is used to determine the date at which a species may become extinct. With that information, those species estimated to be closest to extinction would be chosen for captive breeding.

Can captive breeding cope with the rate at which species are approaching extinction? Rossiter (1990) has claimed that within the next 100 years, about 2000 species of terrestrial vertebrates including 160 primate species, 100 large carnivore species, 100 artiodactyls, 800 bird species and several hundred amphibian and reptile species would have to be captive bred to save them from extinction. There has been opposition to captive breeding of some large mammals such as the Javan rhinoceros (*Rhinoceros sondaicus*). This species, which numbers only 60–70 individuals, is found in only two localities: Ujung Kulon National Park on the south-western tip of Java and in southern Vietnam. Some people have proposed that a captive breeding programme is the only way to save this species but such a programme would require about half the animals to be kept in captivity. On the other hand, it has been argued that the Javan rhinoceros is best conserved in situ and that using this species as a 'flagship species' would attract much funding and help to conserve a natural area of global importance (Mackinnon, 1991).

1.5.2 Protected areas

It seems that, in general, conservation of plants and animals is best done in situ, that is in natural and semi-natural areas. Protected areas of many kinds have a role to play in conserving the world's biodiversity, in monitoring global change and in long-term ecological studies. Some could also be important for sustainable development. For example, Biosphere Reserves are protected areas recognized by the UNESCO Programme on Man and the Biosphere (MAB) for their value in supporting sustainable development and also their value in providing scientific knowledge. There are about 300 Biosphere Reserves in 75 countries ranging in size from 70 million ha (Northeast Greenland Reserve) to 500 ha (North Bull Island in Ireland). There is an important question to be asked here. What criteria are used to select protected areas? That question is addressed in Chapter 6.

1.5.3 Conservation organizations

There are many voluntary conservation organizations around the world and membership has increased dramatically, especially between 1988 and 1990. For example, in Britain, membership (in thousands) for some organizations in 1971, 1981 and 1991 respectively was as follows: Council for the Protection of Rural England (21, 29, 45), Friends of the Earth (not including Scotland: 1, 7, 230), Ramblers Association (22, 37, 85), RSNC, The Wildlife Trusts Partnership (64, 143, 227), Royal Society for the Protection of Birds (98, 441, 852), The World Wide Fund for Nature (12,

60, 261), The Woodland Trust (20 in 1981 and 65 in 1991). These increases are indicative of the growing interest in environmental matters as well as a concern about various environmental issues.

Some major organizations involved in world conservation projects include the International Board for Plant Genetic resources (IBPGR), the World Conservation Union (IUCN), the World Resources Institute (WRI), the World Wide Fund for Nature (WWF), the World Conservation Monitoring Centre (WCMC) and the UNESCO Programme on Man and the Biosphere. The IUCN has its Headquarters in Switzerland but has bases in many countries throughout the World. Both the IUCN and the WWF have long been involved in identifying projects in need of support as well as in financing conservation projects throughout the world. Together, they joined forces with UNEP to prepare and publish the World Conservation Strategy in 1980, a milestone document which promoted three objectives:

1. Maintenance of ecological processes – as life support systems;
2. Preservation of genetic diversity, conservation of wild species;
3. Sustainable utilization of species and ecosystems.

The sequel to the World Conservation Strategy, entitled *Caring for the Earth: A strategy for sustainability* has been prepared by the IUCN, WWF and UNEP and was published in October 1991.

Many countries (52 by 1991) responded to the World Conservation Strategy by preparing a national response and there is no doubt that the World Conservation Strategy has had an impact on policy makers. However, despite the influence of the World Conservation Strategy, some would say that the goals of conservation, resource management and environmental quality are not going to be high priorities for allocation of shrinking resources in developing countries because those countries are too concerned with social and economic problems (Lee, 1989). That is, of course, too simplistic because in fact social and economic problems are interdependent on sustainable use of resources.

Monitoring the status of species and ecosystems is an important aspect of biological conservation (Spellerberg, 1991). For example, the World Conservation Monitoring Centre, which has its origins in the IUCN, undertakes the important task of monitoring the status of species and the status of protected areas (Figure 6.1). The development of databases by organizations such as the WCMC and the US Fish and Wildlife Service (Endangered Species Office) has been central to the success of conservation monitoring organizations. Rapid developments in information technology have greatly improved these databases by providing more powerful and efficient methods of collecting, storing, analysing and communicating large sets of data.

1.5.4 Wildlife law

Wildlife law has had an important role and continues to be an important aspect of conservation. Particularly important international conventions include: the Convention on International Trade in Endangered Species (CITES), the World Heritage Convention and the Ramsar Convention (The Convention on Wetlands of International Importance Especially as Waterfowl Habitat). Others include the Bonn Convention on the conservation of Migratory Species of Wild Fauna, the Convention on the Conservation of Antarctic Marine Living Resources (CCAMLR), and the Berne Convention on the Conservation of European habitats and wildlife. All of them have made contributions to conservation but in some cases that contribution has come only after many years of discussions and negotiations. The Ramasar Convention, for example, had its beginnings at a conference on wetlands held in 1962 (the MAR Conference) in Saintes Maries de la Mer, France. It was not until 9 years later in 1971 that it was signed. The Ramsar Convention finally came into force in 1975, 13 years after the conference in France.

Both the Ramsar Convention and the World Heritage Convention (adopted in 1972 by the General Conference of UNESCO) are two world treaties which deal with conservation of certain habitats and which have similar goals. The Ramsar Convention provides for the conservation of internationally important wetlands whereas the World Heritage Convention provides for the identification and protection of natural (and cultural) areas of outstanding universal value. When a region has been designated a World Heritage Site, it can benefit from international aid (minimum $1 million per year), thanks to the World Heritage Fund.

Another world convention (the 127th) which has attracted the interest of 76 countries is on its way: the International Convention to Conserve Biological Diversity (IUCN, 1990). A fundamental requirement of this convention would be an agreement on a list of centres of biodiversity, i.e. a priority list. If this convention does proceed then there will need to be a judicious appraisal of the method for determining priorities because those centres will then attract relatively high proportions of scarce financial and logistic resources. Useful methods may be those already used by the International Council for Bird Preservation (ICBP) for the Biodiversity Project which identifies Endemic Bird Areas (p. 142).

1.5.5 Communication and collaboration

Biological conservation is one aim of the science of conservation biology and conservation biology is an interdisciplinary subject including aspects of taxonomy, genetics, biogeography, ecology and other disciplines. Clearly, therefore, biological conservation is dependent on many disciplines. For example, plants and animals cannot be conserved if they are not known. There are many organisms yet to be named and there is a continuing need for taxonomic and systematic skills. For example, there is sometimes the problem of deciding what is a legitimate species. Bad taxonomy can be a problem when distinct species are not afforded species status (Gittleman and Pimm, 1991). It is obvious, therefore, that before the questions of criteria and priorities can be addressed fully, we need information; we need more taxonomic studies and we need more databases to store and communicate that information.

Biological conservation is also reliant on surveying and data collection. Access to those data is dependent on clear communication between researchers and practitioners and also on collaborative efforts between conservation organizations and other organizations. This interdisciplinary approach and the importance of collaboration and cooperation was highlighted in McNeely's (1990) account of the IUCN's programme to develop a Global Strategy for Conserving Biological Diversity as follows:

1. The need to define a broader constituency for conservation (conservation has suffered from being seen as a highly specialized sector of only limited relevance to the real needs of society);
2. The need to find ways of bringing local people into the conservation movement;
3. The need to ensure sufficient information is available as a basis for reaching decisions;
4. Better application of science to management;
5. Devising new institutional approaches to conserving biological diversity; for example, how can universities play a greater role?
6. Establishment of priorities for conservation.

1.6 EVALUATION AND ASSESSMENT

1.6.1 Land evaluation

Land evaluation and assessment is a process whereby the potential use or suitability of the land for various purposes or alternatives uses are assessed on the basis of certain criteria. Those uses include building houses, agriculture, forestry, water resources, recreation, landscape conservation and nature conservation. Methods of evaluation have been developed for all these uses and these evaluations are very important for planning the best use of the land and for devising the best management options, particularly because there are so many competing demands on the land.

Evaluation, protection and management of agricultural land is undertaken in many countries throughout the world but there has sometimes been a lack of success in attempts to preserve the best of agricultural land. This lack of success has initiated various attempts to devise better assessment methods. For example in the USA, the location and protection of prime agricultural land has for many years been undertaken by the Soil Conservation Service (SCS). Recently the SCS has devised a Land Evaluation and Site Assessment System (LESA) to enumerate and assess the appropriate physical, social and economic factors so as to meet the planning objectives of the particular region (DeMers, 1985; Luckey and DeMers, 1987). This system is designed to determine the quality of land for agricultural uses and to assess sites or land areas for their agricultural viability. The LESA has two parts; a land evaluation component (based on socioeconomic factors) and a site assessment component (based on soil characteristics). Automation of LESA has recently been undertaken with Geographic Information Systems (GIS) resulting in improved applications of LESA (DeMers, 1989).

1.6.2 Objective and subjective criteria in evaluations and assessments

Assessments of the suitability of land for agriculture and other uses can be based on fairly simple measurements of soil quality, rainfall, irrigation potential and other criteria, most of which can be measured or appraised objectively. By way of contrast, objective criteria such as species richness or rarity have been used alongside subjective value judgements as a basis for evaluating land for use as a national park or a nature reserve.

The use of subjective value judgements has proved to be contentious because of difficulties of assessing the value of a landscape or a piece of nature such as an ancient woodland. So, what is nature worth? A report in an English newspaper in 1991 calculated the value of great crested newts (*Triturus cristatus*) as being £1111.11 each. These newts are protected in Britain and a pond supporting 117 newts was in the path of a new road. The cost of building a new pond for the newts was £130 000. More seriously, willingness to pay is the basis of one approach to valuing biodiversity but does this mean that an ancient woodland, a giant redwood or a desert community is worth what someone or a group of people is prepared to pay for it?

Works of art are only worth what someone is prepared to pay for them but ironically some works of art, such as a painting of a flower or landscape, can sometimes attract sums of money far in excess of what the real thing could fetch (Figure 1.3). Does this mean that we should be less concerned about losses in landscapes because they can be captured forever in a painting? Another example of contrasting views about the value of art and nature comes from some controversy about whether or not works of art can be exported. Such feelings and laws do not seem to apply to 'works of nature' such as highly scientifically important fossils. For example, government officials in Britain have argued that the oldest fossil reptile in the world, does not rank alongside paintings when it comes to granting export licences (Bown, 1991).

1.6.3 Determining priorities for conservation

A common element of many aspects of conservation is the need to identify conservation needs and the need to prioritize those needs (bearing in mind that priority ranking has its critics, see p. 161). For example in ex situ conservation, botanical gardens and zoos would find it impractical to try to conserve all threatened species of plants and animals. Resources dictate that a selection has to be made, the basis of which may depend on the ulterior motives of the zoo or botanical garden (plants and animals which attract visitors). Most zoos have focused their attentions on large vertebrates (especially birds, mammals and reptiles) at the expense of many other taxonomic groups. But is it in the best interests of conservation that: 'Because the world stands to lose a higher proportion of large vertebrates than of other taxa, some disproportionate expenditure on their conservation is justified'? (Reid and Miller, 1989)

Similar questions arise with in situ conservation. Resources and finances are limited and therefore criteria and priorities are necessary for conservation. Although some criteria, especially strictly scientific criteria, may appear to be in conflict with public opinion, the public's perception of what should be conserved can be turned to the advantage of many species. For example, the conservation of well-publicized species (or flagship species), such as the giant panda (*Ailuropoda melanoleuca*), can be used to attract funds for conservation in general and not just for the giant panda, and these funds can be used for the conservation of many species including lesser-known species. Where to start and how to establish priorities are not easy questions to answer but there have been some attempts. For example, three guidelines for setting biodiversity conservation priorities were suggested by Reid and Miller (1989): (1) distinctiveness, (2) utility, (3) threat (Box 1.1). In general terms those criteria are very logical but consideration of the underlying arguments prompts a number of specific questions:

1. If single species are the target of conservation efforts then what are the best criteria for the selection of those species?
2. Saving the last fragments of nature sometimes requires a choice to be made but can that choice be based on any taxonomic or ecological criteria as well as utilitarian and intrinsic features?
3. Should the overall aim be to maximize biodiversity wherever possible or is it better to combine utilitarian and conservation interests?
4. Should there be established criteria for the selection of protected areas?
5. Would it be sensible to direct immediate efforts towards centres of

Box 1.1 Criteria for setting biodiversity conservation priorities. From Reid and Miller (1989) with kind permission of the World Resources Institute

Given that some species and genes are sure to be lost and some natural habitats converted to other land uses in the coming decades, how should scarce human and financial resources be spent to minimize biotic impoverishment and maximize biodiversity's contribution to human well-being? How can the most useful components of biodiversity be saved for current or future use? And how can such ethical considerations as our responsibility to other species and to future generations be incorporated into conservation priorities?

Clearly, these questions cannot be answered with scientific information on the nature, distribution, and status of biodiversity alone. Such factors as awareness of the problem and its solutions, perceived trade-offs with other basic human needs, and local political considerations will all influence conservation priorities. Nonetheless, three rules of thumb derived from the biosciences can help scientists and policy-makers identify the various influential factors and evaluate the trade-offs and value judgments made in setting priorities.

1. Distinctiveness: numbers aren't everything.

To conserve the widest variety of the world's life forms and the complexes in which they occur, the more distinctive elements of this diversity should receive high priority. For example, preserving an assemblage of plants and animals is comparatively more important if those species are found nowhere else or are not included in protected areas. Similarly, if the choice is between conserving a species with many close relatives or a species with few, the more distinctive species should be saved.

Maintaining the highest number of species without considering their taxonomic position makes little sense, as a comparison of marine and terrestrial environments shows. Terrestrial environments contain at least 80% of the world's total species, mainly because vascular plants and insects are so numerous on land – accounting for nearly 72% of all described species in the world – and so poorly represented in marine environments. However, the sea contains greater proportions of higher taxonomic units. Marine ecosystems contain representatives of some 43 phyla while terrestrial environments are home to only 28 phyla. The sea contains fully 90% of all classes and phyla of animals (Ray 1988). Clearly, efforts to conserve the widest array of biodiversity must give attention to all levels of the taxonomic hierarchy.

2. Utility: global or local, current or future?

To some extent, utility is always a criterion in setting biodiversity conservation priorities. Nobody opposed the extinction of wild populations of the smallpox virus, and most people would agree that it's more important to conserve a subspecies of rice than a subspecies of a 'weed.' Similarly,

Box 1.1 continued

maintaining forest cover in the watershed above a village or irrigation project seems more important than preserving a similar forest in a region where increased flooding or soil erosion would not affect human activities downstream.

Of course, many decisions about which species to save are more difficult than the one governing smallpox's fate. In general, questions about the usefulness of species, genes, or ecosystems can be complicated by two factors. One is the issue of inter-generational equity – whether today's needs outweigh those of future generations. It seems reasonable enough to expect future generations to value the species most valued today, but it is impossible to predict humanity's exact needs, and many other species of unknown value today are sure to become more important in the future. Amid attempts to conserve species prized today, some value has to be assigned to future generation's needs.

How useful species, genes, or ecosystems are depends on whether today's needs outweigh those of future generations and who gets the benefits.

Second, we must ask the question: 'Utility for whom?' Biodiversity has obvious values to local communities, nations, and the entire world, but the benefit is not identical for each group. From a global perspective, the conservation of a region's biodiversity might help to regulate climate, influence the atmosphere's chemical composition, and provide all of humanity with industrial products, medicines, and a source of genes for crop breeding. Locally, conservation may also provide people with fuel, clean water, game, timber, aesthetic satisfaction, and important cultural symbols or resources. Because the conservation benefits received by the local and global communities are not congruent, international and local priorities will also differ. To humanity at large, conserving tropical forests matters more than conserving semi-arid deserts since the forests contain a tremendous variety of life and heavily influence global climate. Locally, however, each region's biodiversity is equally valuable since it provides essential ecosystem services that local people rely upon Neither perspective is necessarily the 'correct' view of biodiversity; either – global or local, current or future – reflects an implicit value judgment.

3. Threat: saving the most beleaguered species and ecosystems first.

In general, the threat to biological diversity is influenced by how widespread species are, how common they are over their range, and by such human pressures as harvesting, land conversion, and environment pollution. The threats to species vary dramatically by region, and in each region some ecosystems are more threatened than others. For instance, Central

Box 1.1 continued

> America's tropical rain forests are less threatened than the remaining frag-
> ments of tropical dry forest in that region. In such cases, the dry forests
> should receive the most attention even though the rain forests contain more
> species.
>
> These three factors – distinctiveness, utility, and threat – help make value
> judgments more explicit as conservation priorities are set. If a strictly global
> perspective is adopted, then preserving the most species-rich sites in the
> world seems most important. Indeed, this global perspective has led some
> organizations to focus on conservation in the 'megadiversity' countries –
> Brazil, Colombia, Indonesia, Madagascar, Mexico, Zaire, and others. In
> contrast, from a nationalist perspective, each country's biodiversity is worth
> roughly what every other country's is. Similarly, if today's needs for genes,
> species, or ecosystems are considered paramount, tomorrow's needs can't
> be too.

    endemism, that is those regions which contain significant numbers of
endemic species?
6. As it is impractical to protect all biotic communities from pollution and
    other impacts, what are the most effective methods for identifying the
    most sensitive communities?

    Obviously these questions are relevant at very different spatial scales
but nevertheless they are the central questions for evaluation and
assessment and they are addressed in the following chapters. In general,
an ecological evaluation is undertaken to identify the conservation
needs of a species or the conservation importance of a natural area. An
ecological assessment of a species is undertaken to give detailed
ecological information about that species which may later be used in
management of that species for conservation. An ecological assessment
of a community is undertaken to provide ecological information about
that community which in turn may help to determine the likely effects
of disturbance and pollution on the ecology of an area. Evaluations and
assessments may be based on both subjective judgements and also on
ecological criteria. For example, Shaw and Zube (1980) in a discussion
about the difficulties of trying to define value, suggested that the
difficulties can be partly overcome by distinguishing between economic
measures of value, social and psychological interpretations, and
ecological measures.

1.6.4 The art and science of evaluations and assessments

Evaluation and assessment of nature for conservation has previously been
described and researched by several people including Ratcliffe (1971,
1977), Goldsmith (1975), Peat (1984) and Götmark *et al.* (1986). Several

organizations have also been very much involved in the development of ecological evaluation methods, notably the US Fish and Wildlife Service. A review of a small number of evaluation methods and an introduction to the idea that it is possible to use ecological criteria effectively to evaluate a community or a species was published in 1981 (Spellerberg, 1981) and then later, wildlife conservation evaluation was reviewed by Usher and several colleagues (Usher, 1986). However, there are those who do not support the idea that an evaluation of a species or a community or natural area can be based on ecology. For example, Helliwell (1985) felt that the term 'ecological evaluation' is an example of meaningless jargon, asserting that there can be no such thing as an ecological principle in the sense of something which says what one can or can not do. I do not accept that view and indeed the approach adopted here is to view ecological principles as those principles which are basic to the science of ecology (like principles and laws of physics). For example, energy is lost at each transfer from one trophic level to another. Ecological criteria are similarly basic to the science of ecology and examples include species richness, species diversity, age-class distribution and production. It is these criteria (together with other criteria such as attractiveness) which I believe can usefully be used in the evaluation and assessment of nature.

The aim is not to dwell on conservation philosophy or definitions but to provide material which has a practical use in the conservation of the world's biodiversity. The essence of this book is therefore to undertake an appraisal of the methods of evaluation and assessment of species, biotic communities and natural areas for conservation while paying particular attention to ecological science. There are few other similar books (see Appendix A) but two are worthy of note: one gives a detailed account of ecological evaluations used in North America, Europe and Australia (Usher, 1986) and the other is especially relevant to New Zealand (O'Connor *et al.*, 1990).

1.7 SUMMARY

The growing human population is exploiting biodiversity to an ever-increasing extent and the human race's niche is expanding at the expense of every other species. Although losses in biodiversity have become widely recognized as a serious consequence of human population growth and exploitation and mismanagement of the natural environment, many governments are far from being active in trying to address the serious problems of losses in biodiversity. There are many benefits and functions of biodiversity, some direct and some indirect. Therefore, conservation of biodiversity can be justified in economic terms as well as in moral, aesthetic and intrinsic terms. It is impractical, however, to conserve all aspects of biodiversity to the same extent and therefore it is necessary to evaluate and assess where conservation efforts can best be applied for the optimum effect. Which species to conserve, which biotic community to conserve, where should nature reserves be established, how best can ecology be incorporated in planning and in environmental assessments are questions which are becoming increasingly important for all the human race.

REFERENCES

Abelson, P.H. (1990) Medicine from plants, *Science*, **247**, 513.

Aylward, B. (1991) The economic value of ecosystems: 3, biological diversity. London Environmental Economics Centre, Gatekeeper series No. LEEC GK, 91–3.

Bown, W. (1991) Export controls on 'works of nature' rejected. *New Scientist*, **131**, 7.

Carson, R. (1962) *Silent Spring*. Houghton Mifflin, Boston.

Cronk, Q. (1988) *Biodiversity. The key role of plants*. IUCN and WWF.

DeMers, M.N. (1985) The formulation of a rule-based GIS framework for county land use planning. *Modeling and Simulation*, **16**, 93–7.

DeMers, M.N. (1989) Knowledge acquisition for GIS automation of the SCS LESA Model: an empirical study. *A1 Applications*, **3**, 12–22.

Dodd, C.K. (1990) Effects of habitat fragmentation on a stream dwelling species, the flattened musk turtle *Sternotherus depressus*. *Biological Conservation*, **54**, 33–45.

Ehrenfeld, D.W. (1976) The conservation of non-resources. *American Scientist*, **64**, 648–56.

Ehrenfeld, D. (1981) *The Arrogance of Humanism*. Oxford University Press, New York.

Erwin, T.L. (1982) Tropical forests: their richness in coleoptera and other arthropod species. *Coleopterist's Bulletin*, **36**, 74–5.

Erwin, T.L. (1991) How many species are there?: revisited. *Conservation Biology*, **5**, 330–3.

Farnsworth, N.R. and Soejarto, D.D. (1985) Potential consequence of plant extinction in the United States on the current and future availability of prescription drugs. *Economic Botany*, **39**, 231–40.

Fearnside, P.M. (1990) The rate and extent of deforestation in Brazilian Amazonia. *Environmental Conservation*, **17**, 213–26.

Gaston, K.J. (1991) The magnitude of global insect species richness. *Conservation Biology*, **5**, 283–96.

Gentry, A.H. (1988) Tree species richness of upper Amazonian forests. *Proceedings of the National Academy USA*, **85**, 156–9.

Gittleman, J.L. and Pimm, S.L. (1991) Crying wolf in America. *Nature*, **351**, 524–5.

Goldsmith, F.B. (1975) The evaluation of ecological resources in the countryside for conservation purposes. *Biological Conservation*, **8**, 89–96.

Götmark, F., Ahlund, M. and Eriksson, M.O.G. (1986) Are indices reliable for assessing conservation value of natural areas? An avian case study. *Biological Conservation*, **38**, 55–73.

Helliwell, D.R. (1985) *Planning for Nature Conservation*. Packard Publishing, Chichester.

Iltis, H.H., Doebley, J.F., Guzman, R. and Pazy, B. (1979) *Zea diploperennis* (Gramineae), a new teosinte from Mexico. *Science*, **203**, 186–8.

IUCN (1980) *World Conservation Strategy: Living Resource Conservation for sustainable Development*. IUCN-UNEP-WWF, Gland.

IUCN (1990) The proposed international convention to conserve biological diversity. Paper submitted to the 15th Session of the UNEP Governing Council. IUCN, Gland.

Kuliopulos, H. (1990) Amazonian biodiversity. *Science*, **248**, 1305.

Lande, R. (1988) Genetics and demography in biological conservation. *Science*, **241**, 1455–60.

Lavigne, D. (1991) Your money or your genotype. *BBC Wildlife*, **9**, 204–5.

Lee, J. A. (1989) Conservation in a World in search of a future, in *Conservation for the Twenty-first Century* (eds D. Western and M. Pearl) Oxford University Press, Oxford, pp. 284–8.

Leopold, A. (1933) *Game Management*, C. Scribner, New York and London.

Lewis, J. R. (1978) The implications of community structure for benthic monitoring studies. *Marine Pollution Bulletin*, **9**, 64–7.

Lewis, J.R. (1990) Rainforest rundown. *BBC Wildlife*, **8**, 386–93.

Lovejoy, T.E. (1980) Tomorrow's ark: by invitation only. *International Zoo Yearbook*, **20**, 181–3.

Lovejoy, T.E. (1984) Application of ecological theory to conservation planning, in *Ecology in Practice* (eds F. di Castri, F.W.G. Baker and M. Hadley) Tycooly International Publishers, Dublin, pp. 402–13.

Lovejoy, T.E. (1989) Deforestation and the extinction of species, in *Changing the Global Environment. Perspectives on Human Involvement* (eds D.B. Botkin, M.F. Caswell, J.E. Estes and A.A. Orio) Academic Press, London, pp. 89–98.

Lovejoy, T.E., Bierregaard, R.O., Rylands, A.B., Malcolm, J.R., Quintela, C.E., Harper, L.H., Brown, K.S., Powell, A.H., Powell, G.V.N., Schubart, H.O.R. and Hays, M.B. (1986) Edge and other effects of isolation on Amazon forest fragments, in *Conservation Biology: the Science of Scarcity and Diversity* (ed. M.E. Soule) Sinauer, Massachusetts, pp. 257–85.

Lubchenco, J., Olson, A.M., Brubaker, L.B., Carpenter, S.R., Holland, M.M., Hubbell, S.P., Levin, S.A., MacMahon, J.A., Matson, P.A., Melillo, J.M., Mooney, A.H., Peterson, C.H., Pulliam, H.R., Real, L.A., Regal, P.H. and Risser, P.G. (1991) The sustainable biosphere initiative: an ecological research agenda. *Ecology*, **72**, 371–412.

Luckey, D. and DeMers, M.N. (1987) Comparative analysis of land evaluation systems for Douglas County. *Journal of Environmental Systems*, **16**, 259–78.

Mackinnon, K. (1991) How about letting then breed by themselves? *BBC Wildlife*, **9**, 454–5.

Marsh, G.P. (1864) *Man and Nature; or Physical Geography as Modified by Human Action*. Scribers, New York; Sampson and Low, London.

May, R.M. (1991) A fondness for fungi. *Nature*, **352**, 475–6.

McNeely, J.A. (1988) Economics and biological diversity; developing and using economic incentives to conserve biological resources. IUCN, Gland.

McNeely, J.A. (1990) A global strategy for conserving biological resources. *Species*, **13–14**, 6–9.

Myers, N. (1979) *The Sinking Ark*. Pergamon, Oxford.

Myers, N. (1983) A priority-ranking strategy for threatened species? *The Environmentalist*, **3**, 97–120.

Myers, N. (1989) *Deforestation Rates in Tropical Forests and Their Climatic Implications*. Friends of the Earth, London.

Nelson, R. and Holben, B. (1986) Identifying deforestation in Brazil using multiresolution satellite data. *International Journal of Remote Sensing*, **3**, 429–48.

O'Conner, K.F., Overmars, F.B. and Ralston, M.M. (1990) *Land Evaluation for Nature Conservation*. Lincoln University Centre for Resource Management, New Zealand.

O'Hara, K.J., Iudicello, S. and Bierce, R. (1988) *A Citizen's Guide to Plastics in the Ocean; More than a Litter Problem*. Centre for Marine Conservation, Washington.

Paine, R.T. (1969) The *Piaster-Tegula* interaction: prey patches, predator food preferences and intertidal community structure. *Ecology*, **50**, 950–61.

Peat, J.R. (1984) A partially objective method for the ecological evaluation of biological communities. PhD Thesis, University of Southampton.

Peters, C.M., Gentry, A.H. and Mendelsohn, R.O. (1989) Valuation of an Amazonian rainforest. *Nature*, **339**, 655–6.

Plotkin, M. (1988) The outlook for new agricultural and industrial products from the tropics, in *Biodiversity* (eds E.O. Wilson and F.M. Peter) National Academy Press, Washington, pp. 106–16.

Prescott-Allen, R. and Prescott-Allen, C. (1978) *Sourcebook for a World Conservation Strategy: Threatened Vertebrates*. General Assembly Paper GA 78/10 Addendum 6, Gland: IUCN.

Prescott-Allen, R. and Prescott-Allen, C. (1991) How many plants feed the World. *Conservation Biology*, **4**, 365–74.

Ratcliffe, D.A. (1971) Criteria for the selection of nature reserves. *Advancement of Science, London*, **27**, 294–6.

Ratcliffe, D.A. (ed) (1977) *A Nature Conservation Review*, Vols. 1 and 2, Cambridge University Press, Cambridge.

Ratcliffe, D.A. (1980) *The Peregrine Falcon*. T. & A.D. Poyser, Calton.

Reid, W.V. and Miller, K.R. (1989) *Keeping Options Alive: The Scientific Basis for Conserving Biodiversity*. World Resources Institute, New York.

Rossiter, A. (1990) Ecology, environment and economics: a pandoran perspective. *Physiology and Ecology Japan*, **27**, 169–89.

Schonewald-Cox, C., Chambers, S.M., MacBryde, B. and Thomas, W.L. (1983) *Genetics and Conservation*. Benjamin/Cummings, London.

Schultes, R.E. and Raffauf, R.F. (1990) *The Healing Forest. Medicinal and Toxic Plants of the Northwest Amazonia*. Dioscorides Press, Portland, OR.

Seal, U.S. (1985) The realities of preserving species in captivity, in *Animal Extinctions* (ed. R.J. Hoage) Smithsonian Institution, Washington DC, pp. 71–95.

Seal, U.S. (1988) Intensive technology in the care of ex situ populations of vanishing species, in *Biodiversity* (eds E.O. Wilson and F.M. Peter) National Academy Press, Washington, DC, pp. 289–95.

Shaw, W.S. and Zube, E.H. (1980) *Wildlife Values.* Centre for Assessment of Non-commodity Natural Resource Values, report No. Tucson, University of Arizona.

Smart, J. (1990) *Grain Legumes. Evolution and Genetic Resources.* Cambridge University Press, Cambridge.

Soule, M.E., Bolger, D.T., Alberts, A.C., Wright, J., Sorice, M. and Hills, S. (1988) Reconstruction dynamics of rapid extinctions of chaparral-requiring birds in urban habitat islands, in *Conservation Biology: An Evolutionary-Ecological Perspective.* Sinauer, Masschusetts.

Spellerberg, I.F. (1981) *Ecological Evaluation for Conservation.* Edward Arnold, London.

Spellerberg, I.F. (1991) *Monitoring Ecological Change.* Cambridge University Press, Cambridge.

Spellerberg, I.F. and Hardes, S.R. (1992) *Biological Conservation.* Cambridge University Press, Cambridge.

Tatum, L.A. (1971) The southern corn leaf blight epidemic. *Science,* **171,** 1113–16.

Usher, M.B. (1986) *Wildlife conservation evaluation.* Chapman & Hall, London.

Wallace, N. (1985) Debris entanglement in the marine environment: a review, in *Proceedings of the Workshop on the Fate and Impact of Marine Debris,* 27–9 November 1984, Honolulu, Hawaii (eds R.S. Shomura and H.O. Yoshia) US Dept. of Commerce, NOAA Technical Memorandum NMFS.

WCED (1987) *Our Common Future.* Oxford University Press, Oxford, New York.

Wilson, E.O. (1988) *Biodiversity.* National Academy Press, Washington.

Winpenny, J.T. (1991) *Values for the Environment.* HMSO, London.

World Bank (1988) *Wildlands: Their Protection and Management in Economic Development.* World Bank, Washington.

WRI/IIED (1987) *World Resources 1987.* Basic Books, New York.

An ecological basis for evaluation and assessment

2.1 INTRODUCTION

Ecology is the scientific study of the interactions between plants and animals and their environment. One important aspect of ecology is the study of the distribution of plants and animals in habitats (spatial distribution) and change in population over time (temporal distribution). Some basic questions in ecology are therefore as follows: what species are present? how many are there? how are they distributed? and what environmental factors determined that distribution? The environmental factors include physical and chemical factors, the soil, plants and animals.

The aim of this chapter is to introduce some basic elements of ecology relevant to ecological evaluation and assessment of nature. The most commonly used criteria used for evaluation for conservation have been diversity, rarity and area (Margules and Usher, 1981; Smith and Theberge, 1986). Therefore, particular attention is given here to the ecological basis of diversity, rarity (species rarity) and area. The criteria naturalness, typicalness and representativeness are also considered as they have been used extensively in assessing areas for nature reserves.

2.2 POPULATIONS, DISTRIBUTION AND ABUNDANCE

One definition of a population is a collection of individuals (either plants or animals) of the same species in a prescribed area. In an evaluation we could be recording a population of lizards on a heathland or a population of oak trees in a mixed deciduous woodland. Population size is simply the number of individuals in that population but in any population there will be different age groups. Measuring the size of a population is not always easy because many organisms are not easily detected or at least it may not be possible to detect the whole population. Whereas in a woodland it may be easy to count the individual trees of various species, counting the individuals of species of birds, butterflies or bats would be more difficult.

In some cases it is necessary, therefore, to sample the population to provide an estimate of the total numbers of that species in the habitat. Providing the same sampling techniques are used, such samples can sometimes be used to compare one site with another.

All organisms have evolved to exist in different environmental conditions, some existing in or tolerant of wide-ranging conditions (euryoecious) and some tolerant of very narrow conditions (stenoecious). The latter are sometimes good biological indicators (p. 52), that is indicators of certain environmental conditions. Those environmental conditions include physical and chemical factors, soil conditions and most importantly other organisms. Different species occur in different locations and the distribution of a species in space is a product of evolution and mankind's interference with the pattern and process of nature.

The distribution of organisms in time as well as space has important implications for data collection. Different species are present or active at certain times and an organism may be present but not visible. The distribution in time applies to plants as well as animals. It is easy to imagine surveys of woodland flora being taken at two different times and resulting in different species lists. That is, not all plants have vegetative growth which is visible all year long. Animals are more varied in their distribution and types of activity in time. For example, the annual migration of some bird species could lead to quite different species lists in the same place but on two different occasions. Some animals overwinter or hibernate and evidence of their presence may be detected only at certain specific times of the year or in a particular season. Many invertebrates are active not only at different times of the year but at different times of the day or night (Figure 2.1). Even the sexes may have different periods of activity. Furthermore, some animals appear in different forms (in insects, metamorphosis from the larva or caterpillar to the winged adult) and at different times of the year some animals show different physiological states (e.g. breeding condition). It is important therefore to bear in mind the temporal aspects of plant and animal activity and condition if an ecological evaluation is based in part on species richness or species diversity. The time (time of day and time of year) of data collection or sampling must be made known.

2.3 THE RARITY OF SPECIES

2.3.1 The concepts

Rarity, threatened and endangered are in common use in conservation texts and not surprisingly there have been many attempts to define these categories (Fitter and Fitter, 1987 give a very good review). From a strictly ecological point of view and in terms of relative abundance, most species are rare and few are common. That is, if we were to record the abundance

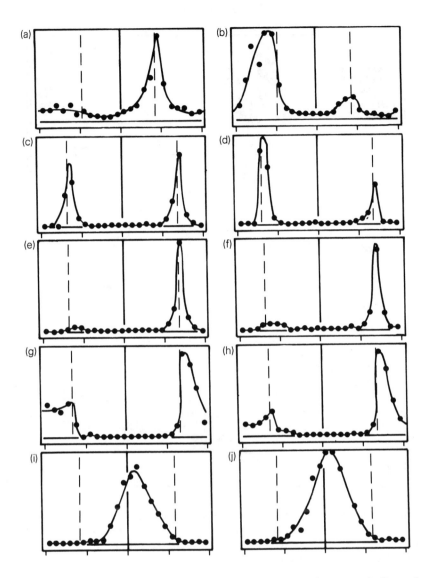

Figure 2.1 Activity curves for different insects in Britain. The curves indicate the abundance (thus activity) of insects of different sexes over 24 h. (a) and (b) female and male moths (*Lithocolletis messaniella*); (c) and (d) and also (e) and (f) female and male flies (*Drosophila* sp.); (g) and (h) female and male mosquitoes (*Culex pipiens*); (i) and (j) female and male cabbage white fly (*Aleyrodes proletella*). 12.00 midday is indicated by the central thick line. From Lewis and Taylor (1964) with kind permission of Professor R.L. Taylor.

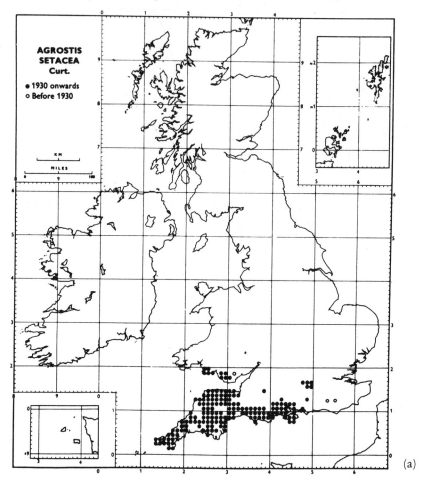

(a)

Figure 2.2 Examples of distribution maps from the *Atlas of British Flora* (Perring and Walters, 1962), reproduced with kind permission of the Botanical Society of the British Isles. Each dot represents one or more records in a 10 km square. The grass *Agrostis setacea* in (a) could be considered to be rare in England and Wales because it is found only in the southwest. On the other hand *Ranunculus lingua* in (b) could be considered rare on a local basis.

of many species we would see that some cosmopolitan species, such as starlings (*Sturnus vulgaris*), bracken (*Pteridium*), reeds (*Phragmites*) or glasswort (*Salicornia*), are represented by many individuals whereas most species are represented by fewer individuals. That being the case, what is meant by rarity?

Species rarity is perceived in various ways and can be expressed in a number of interesting ways. What is often meant as a rare species is one

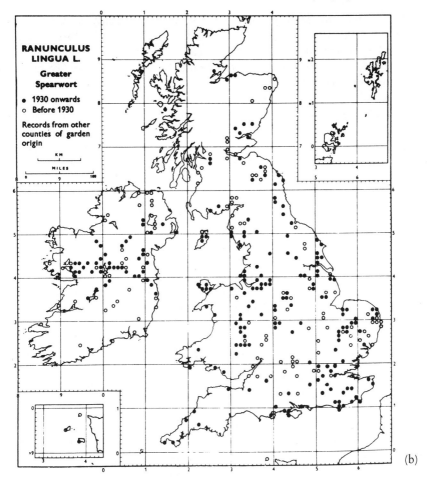

(b)

which is not often encountered: thus rarity can be a subjective assessment of how often a species is thought to be encountered. If a species is protected then it may also be thought of as rare. From an ecological point of view we can consider rarity as being a quantifiable expression of the abundance and distribution of a species. Diminishing numbers over time may lead to a species being classed as rare and indeed many species in Red Data Books are included for that reason (Table 4.3).

The abundance of a species (shown by a distribution map) is one popular way of assessing rarity. Distribution maps can be of different scales and on a countrywide basis a standard way of recording distribution has been to use units of 10 km × 10 km based on the national grid. For example the two distribution maps in Figure 2.2 are based on 10 km squares. A dot in a square means that the species has been recorded at least once somewhere within that square. In Figure 2.2, there is a similar number of

10 km squares for each species but one species occurs only in the southwest and the other species has a fragmented distribution. The former species could be considered nationally rare because of a limited distribution in one part of the country and the latter could be considered locally rare but common on a countrywide basis.

The use of 10 km squares for mapping distributions has proved to be very useful and has been used as a basis for quantifying rarity and for calculating the degree of threat faced by a species (p. 97). Distribution maps using 2 km squares (tetrads) can give a more complete picture but of course the amount of data required increases. The change in the number and distribution of records can be used to monitor a species but records on a 10 km square could be misleading. For example, imagine that a hypothetical species has been recorded in 23 10 km squares (Figure 2.3, grid 1). The same species has also been mapped on the basis of tetrads; the tetrads in the lower part of Figure 2.3 represent the same area in each of the four grids above. A few years later the distribution is examined again and in this hypothetical example we could consider three alternative results (Figure 2.3, grids 2,3 and 4). The decline in number of 10 km squares is further qualified by looking at the change in tetrads with the following results:

Grid no.	10 km squares	Tetrads
2	15	22
3	16	54
4	17	24

On the basis of change in number of 10 km squares there appears to be no difference but the change in number of tetrads reveals a marked difference with grids 2 and 4 showing the greatest decline. The grid 4 has one other worrying change in distribution and that is the distribution records following what appears to have been a diagonal, linear feature (from top left to bottom right) have gone. This indicated that there has been more fragmentation than in grid 2.

The various forms of rarity have been classified in a number of ways and one particularly good classification was described by Rabinowitz (1981). She derived seven forms of rarity based on three attributes: geographical range, habitat specificity and local population size (Table 2.1). If each of these three attributes is dichotomized then an eight-celled block emerges. Disregarding the top left cell (species with wide ranges and several habitats and locally high abundances) leaves seven forms of rarity. This classification is a good, thought-provoking introduction to the taxonomic and ecological basis of rarity. However, in practical terms, distribution maps showing abundance on the basis of a certain scale (for example, 10 × 10 km square) is a good, basic way or quantifying rarity.

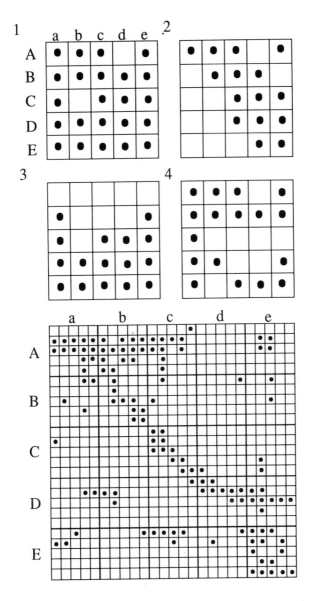

Figure 2.3 The distribution of a hypothetical species is shown in grid 1 on the basis of 10 km squares and the same distribution is shown below but on the basis of 2 km squares (tetrads). The grids 2,3 and 4 are different versions of grid 1 and the large grid below is an enlargement of grid 1. A change in distribution from grid 1 to grids 2, 3 or 4 reveals little difference in the loss of 10 km squares. However, the loss on tetrads reveals important differences: grids 2 and 4 have lost more tetrads than grid 3. However, grid 4 is the worst because of the increase in isolation (loss of tetrads along a linear feature).

Table 2.1 A typology of rare species based on three characteristics: geographic range, habitat specificity, and local population size based on an example from the North American Flora. From Rabinowitz (1981).

Geographic range	Large		Small	
Habitat specificity	Wide	Narrow	Wide	Narrow
Local population size Large, dominant somewhere	Common. Locally abundant over a large range in several habitats e.g. *Chenopodium album*	Locally abundant over a large range in a specific habitat e.g. *Rhizophora mangle*	Locally abundant in several habitats but restricted geographically e.g. *Cupressus pygmaea*	Locally abundant in a specific habitat but restricted geographically e.g. *Shortia galacifolia*
Small, non-dominant	Constantly sparse over a large range and in several habitats e.g. *Setaria geniculata*	Constantly sparse in a specific habitat but over a large range e.g. *Taxus canadensis*	Constantly sparse and geographically restricted in several habitats. Non-existent?	Constantly sparse and geographically restricted in a specific habitat e.g. *Torreya taxifolia*

2.3.2 Rarity indices

In addition to the popular use of abundance and distribution maps as a basis for qualifying what is meant by rarity, some authors have devised rarity indices. For example, rarity indices calculated from species distribution data of water and ground beetles have been developed by Luff *et al.* (1989) for use in the assessment of conservation values of sites in northeast England. The indices were calculated in three steps and were based on distribution records (in 2 × 2 km or tetrads) for each species of beetle in the north-east. One scale of rarity was similar to that used by Dony and Denholm (see p. 119) where species were recorded in 1, 2–3, 4–7, 8–15, 16–31, 32–63 and 64 or more tetrads and were given scores of 7 (1 tetrad) to 1 (64 or more). A second scale was similar except that it was scored geometrically, that is 64, 32, 16, 8, 4, 2 and 1. A sum of the scores gave the species rarity total (SRT). A third index, the rarity association was calculated using species that scored 2 or more in the geometric scale. So as to eliminate bias caused by one very rare species in a list, the highest score was reduced to that of the next nearest score. That is, if a species list contained scores of 32, 8, 4 and 1 then the 32 would be reduced to 8 and all the scores of two or more summed. The rarity association value for the site would therefore be 20. This score would then be added to the SRT to give a rarity association total. With all indices, the totals were divided by the species richness (number of species) so that the quality of sites could be compared on a unit basis.

2.4 SPECIES COMPOSITION, RICHNESS AND DIVERSITY

2.4.1 Species composition

In addition to habitat diversity (meaning the variety or number of habitats), common elements in many ecological evaluations are species composition, species richness and species diversity. Unfortunately these very useful terms can sometimes be used in a less than rigorous fashion. The species composition is simply the species present (sometimes abundance of each species might also be recorded). In Tables 2.2 and 2.3 the species composition of

Table 2.2 Beetles sampled by pitfall trapping on a remnant 108 ha piece of heathland in Dorset, UK. Data from a student project, University of Southampton (H.T.M. Winter)

Family	Species	Number
Carabidae	*Pterostichus cupreus*	5
Carabidae	*P. nigrita*	1
Chrysomelidae	*Lochmaea suturalis*	92
Curculionidae	*Sitona humeralis*	1
Carabidae	*Dyschirius globosus*	4
Carabidae	*Bembidion guttula*	1
Leiodidae	*Catops fuliginosus*	1
Staphylinidae	*Anthobium* sp.	1
Staphylinidae	*Atheta fungi*	1
Carabidae	*Amara nitida*	1
Carabidae	*Loricera pilicornis*	1
Staphylinidae	*Philonthus varius*	1
Staphylinidae	*Xantholinus linearis*	1
Curculionidae	*Strophosomus sus*	6
Curculionidae	*S. curvipes*	2
Curculionidae	*Micrelus ericae*	4
Carabidae	*Carabus violaceus*	3
Staphylinidae	*Olophrum piceum*	1
Scarabaeidae	*Onthophagus fracticornis*	1
Cryptophagidae	*Cryptophagus setulosus*	1
	Total	129

Species richness = 20
Species diversity in the form $1 - \Sigma Pi^2 = 0.48$

the beetles is shown by the list of species but note that in Table 2.2 the composition is dominated by one species (*Lochmaea suturalis*). The term species richness should strictly be used for the total number of species. In Table 2.2 the species richness of the beetles is 20 compared to a greater species richness of 29 shown in Table 2.3.

Table 2.3 Beetles sampled by pitfall trapping on a remnant 11 ha piece of heathland in Dorset England. Data from a student project, The University of Southampton (H.T.M. Winter)

Family	Species	Number
Curculionidae	*Strophosomus sus*	2
Curculionidae	*S. curvipes*	3
Chrysomelidae	*Lochmaea suturalis*	11
Curculionidae	*Acalles ptinoides*	2
Scydmaenidae	*Cephennium gallicum*	4
Carabidae	*Carabus violaceus*	4
Carabidae	*C. problematicus*	1
Carabidae	*Bembidion lampros*	3
Silphidae	*Choleva agilis*	3
Scolytidae	*Hylastes angustatus*	1
Hydrophilidae	*Cryptopleurum* sp.	1
Nitidulidae	*Epuraea* sp.	1
Carabidae	*Amara aenea*	1
Carabidae	*Acupalpus meridianus*	2
Lathridiidae	*Lathridius minutus*	1
Tenebrionidae	*Cylindrinotus laevioctostriatus*	1
Carabidae	*Notiophilus biguttatus*	1
Staphylinidae	*Cypha longicornis*	1
Staphylinidae	*Atheta fungi*	2
Staphylinidae	*Xantholinus* sp.	2
Staphylinidae	*Tachyporus chrysomelinus*	3
Staphylinidae	*Amischa analis*	1
Staphylinidae	*Bolitobius cingulatus*	1
Staphylinidae	*Olophrum piceum*	17
Staphylinidae	*Anthobium atrocephalum*	4
Staphylinidae	*Mycetoporus lepidus*	1
Staphylinidae	*Philonthus cognatus*	7
Staphylinidae	*Atheta* sp.	1
Staphylinidae	*Stenus impressus*	1
	TOTAL	83

Species richness = 29
Species diversity in the form $1 - \Sigma Pi^2 = 0.92$

2.4.2 Species richness

Species richness is simply the number of species but some authors have used a species richness index based on the total number of species and total number of individuals. Species diversity is an expression often used when what is really meant is species richness. Diversity is a measure of the relative abundance of species or the way in which each species is represented by individuals. For example in a hypothetical example of five species A,B,C,D and E represented by a grand total of 100 individuals there could be two extremes of diversity. In one extreme each species could be represented by 20 individuals (highest diversity) and in the other extreme one species could be represented by 96 individuals and all the others by one individual each (lowest diversity).

Both species richness and species diversity can be expressed as an index and indeed there are many indices of species richness and diversity (Spellerbeg, 1991). For example two indices of species richness are as follows

$$SR = \frac{S-1}{\log N} \quad \text{(Margalef, 1951)}$$

$$SR = \frac{S}{\sqrt{N}} \quad \text{(Menhinick, 1964)}$$

Where SR is the index of species richness, S is the total number of species in the sample or area and N is the total number of individuals. The former is particularly useful when there are very large numbers of individuals.

2.4.3 Species diversity

Three commonly used indices of species diversity are Simpson's, Shannon–Weiner's and the Alpha Diversity Index. The first of these may be expressed as follows:

$$D = \sum_{i=1}^{S} P_i^2 \quad \text{(Simpson, 1949)}$$

The Shannon-Weiner Index of diversity is expressed as

$$D = - \sum_{i=1}^{S} P_i(\log_e P_i) \quad \text{(Pielou, 1966)}$$

where P_i is the proportion of the ith species. The proportions are given by

$P_i = n_i/N$

In both cases of species diversity index, P_i is the proportion of individuals of species i in the sample or area being sampled. For example if there were two species in a sample represented by 99 and 1 individuals, then Simpson's Index of diversity would be expressed as follows:

$$D = \Sigma \ (0.9^2 + 0.01^2)$$
$$= 0.81$$

By way of contrast, if a sample had two species which were each represented by 50 individuals then the diversity index (Simpson's) would be 0.5.

The Alpha Diversity Index (Fisher *et al.*, 1943) is expressed as follows:

$$S = \alpha \ \text{Log}_e \ (1 + N/\alpha)$$

where S is the total number of species, N is the total number of individuals and α is the diversity index. The value for α is read from a nomogram (see Southwood, 1966; Spellerberg, 1991).

While an index of diversity can be a convenient way of summing a lot of information, it is just a mathematical expression. It should be put into context, that is the minimum and maximum values possible should be made known. Sample size can affect the index but less so in the case of the Shannon-Weiner Index. Simpson's Index gives relatively little weight to rare species and more emphasis to the common species. For a more detailed account of species richness indices and diversity indices see Magurran (1988) or Spellerberg (1991).

In Tables 2.2 and 2.3 the data for an ecological evaluation (based on the beetles only) can be summarized as follows. For the 108 ha of heathland, the species composition is dominated by one species, the species richness is 20 and the species diversity is 0.48 (Shannon-Weiner); for the 11 ha of heathland, the species composition is represented over 14 families and there is no one species which is far more common than any other, the species richness is 29 and the species diversity is 0.92 (Shannon-Weiner). The latter heathland fragment was surrounded by a conifer plantation and the 108 ha fragment was an isolated common (urban common ground or 'park') near a small village.

2.5 AREA AND SPECIES RICHNESS

The extent of an area being evaluated is a criterion commonly used in ecological evaluations, not least because it has important implications for conservation. Of the many aspects which could be discussed, two are worthy of particular note: minimum area and species–area relationships. The process of reduction, fragmentation and isolation of natural areas is a common theme throughout this book and it is important therefore to emphasize the ecological implications. As natural areas become smaller and smaller, it must follow that some species will eventually no longer be able to find all their resources (food, breeding sites etc.) in that reduced area.

Changes in land use, reduction and fragmentation of natural areas make this process more complicated with regard to selection of natural areas for nature reserves. For example, a small area of woodland and scrubland surrounded by managed grassland by itself may be considered too small to support certain species of woodland birds. When the woodland and scrubland is considered in context of the wider landscape, a greater number of species of birds could possibly be supported if some trees and scrubland were allowed to grow between the nature reserve and other woodlands and hedges in the nearby countryside. This landscape ecology approach is of course dependent on having some control over the management of the land around the nature reserve.

The idea that extent or area is an important criterion has been much influenced by the common observation that more species are found in larger areas: as area increases the number of species (of some taxonomic groups such as woodland birds) has been found to increase. There are, however, exceptions, notably the number of beetles in lowland heathland fragments in England (Tables 2.2, 2.3; see also Webb, 1986). For those plant or animal groups in which the species richness does show an increase with area, in general it has been found that the number of species doubles with a tenfold increase in area. This relationship has been an incentive for vigorous research, much of it sadly being influenced by muddled interpretation and application of island biogeographical theory (Spellerberg, 1991); that is, the ecology of fragments of woodlands and other ecosystems has sometimes been likened to the ecology of oceanic islands. There are various theories which attempt to explain the species richness and rates of colonization of different kinds of islands (small and large, remote and close to continents) and it is those theories which have been wrongly applied to fragments of terrestrial ecosystems.

Although it can be shown that for some plant or animal groups, larger fragments do support more species, there are other studies which show that an equally large number of species can be conserved by designating several small areas amounting to the same area as a single large fragment. This applies not only to terrestrial habitats but also to aquatic habitats. For example, Helliwell (1983) examined vascular plant species at 101 open water sites in England to test the hypothesis that large sites are more valuable than small sites. The hypothesis was supported by the data but a large amount of variation was due to reasons other than area. Not surprisingly, some small ponds were found to be of greater conservation value than ponds ten times their size. The lesson to be learnt is that in an evaluation of an area, habitat or community, one must be alert to exceptions and must take note of intrinsic values.

2.6 COMMUNITIES AND COMMUNITY STRUCTURE

2.6.1 Communities

A community (biological or biotic community) is a group of organisms belonging to different species that are found in the same habitat or area and which interact by way of energy transfer through trophic levels. Sometimes communities are classified on the basis of a dominant taxonomic group or species: for example, mixed deciduous woodland or chalk grassland. The way in which communities function has been a central theme of ecological research for many years. The flow and cycling of energy, nutrients and minerals in communities and the interrelationships between species have been focal points for ecology.

The ecology of some species can be linked closely with many other species to the extent that those few species are 'key' elements of the community: thus the concept of key or keystone species. Some carnivores, termites, insect and bird pollinators and seed dispersers and many plants which provide resources for many other species have been identified as keystone species. The concept of keystone species is mentioned (p. 81) in relation to the rationale underlying species evaluation.

The structure of an assemblage of species, that is the way in which species are distributed in a community, can be described in a number of ways using ecological parameters such as the species composition, relative abundance and species distribution patterns. It is those parameters which may be used as a basis for qualifying what is meant by naturalness, typicalness and representativeness. The two criteria, naturalness and typicalness are two of 10 criteria which are used widely for selection of areas as nature reserves. Naturalness is easy to define and refers to intactness or integrity of ecosystems and processes free of human influence (e.g. Wright, 1990). Typicalness is more difficult to define but essentially it refers to those areas which are selected on the basis of their characteristic and common habitats, communities and species.

2.6.2 Quantifying naturalness, typicalness and representativeness

Anderson (1991) has described a conceptual framework for evaluating and quantifying naturalness or integrity of ecosystems. Three indices of naturalness were proposed: the degree to which the system would change if humans were removed, the amount of cultural energy required to maintain the functioning of the ecosystem as it currently exists (that is, auxiliary energy derived from energy sources (hydroelectric nuclear, tidal) and managed and used by human cultures), and the complement of native species currently in an area compared with the group of species in the area prior to settlement. The second index is based on the observation that a natural system will function without subsidies of cultural energy. The amount of cultural energy required to maintain day to day functions in the ecosystem could provide a measure of naturalness.

Measurements of typicalness have been based on the study of species assemblages: that is, we ask what species typically occur together? If species A,B,C,D and F typically occur together in one area and species A,C,E and F occur typically together in another area, then we can use these two groups (assemblages) of species as a basis for defining habitats. Phytosociologists for example have used this method to define habitats on the basis of the plant species composition (see for example Kershaw and Looney, 1985). More recently, there has been research on the use of certain groups of insects to define habitats, and research which makes combined use of ordination and classification to handle large data sets.

With small data sets, indices of similarity can be used as a basis for identifying species assemblages and classifying groups of species. Classification by way of cluster analysis (Sneath and Sokal, 1973) is a method which provides a useful way of examining data sets. There are many kinds of cluster analyses and one good example is that used by Savage (1982) in a study of water boatmen (Corixidae) in Britain. Savage used indices of similarity to construct a dendogram to show the affinities between water boatmen in a series of lakes. He was able to classify the groups of water boatmen and therefore describe typical species assemblages for the different lakes. Another example using water beetles for evaluating and ranking wetland sites is given in section 5.3.

Representativeness can be thought of in two ways. First, as a series of natures reserves which represent the main types of communities in a country. For example, in New Zealand, representativeness, has been endorsed as a basis for nature conservation policy (O'Connor *et al.*, 1990) because at the moment the protected area network is thought not to be representative. Second, in large protected areas where there may be different types of plant communities, the concept of representativeness is based on the notion that there should be a range of species and habitats which represent the range of variation found within a defined land class or region. This has been used in international programmes such as the Biosphere Reserves Programme (UNESCO, 1974) and the World Conservation Strategy (p. 24). In Australia, representativeness has been widely used in evaluations and assessments.

It is because representativeness cannot be measured directly that there are several measurements associated with it. Ecological surveys and classification of species assemblages (at different scales, national to regional) are keys to studying representativeness. In Australia, for example, representativeness has previously been assessed using vegetation maps and also an hierarchial system of vegetation types based on structural formations (Specht *et al.*, 1974). Classification can be based in part on species associations as outlined in section 2.8.

2.7 BIOLOGICAL INDICATORS AND INDICES

2.7.1 Indicators

Indicator species include those species which are sensitive to physical and chemical conditions. Generally speaking it is the presence and change in abundance of those species which indicate pollution and disturbance. Some species of lichens, for example, are known to be sensitive to SO_2 pollution. The use of lichens as bioindicators of air-borne pollution has been well documented for the last 130 years (Hawksworth and Rose, 1976).

As well as individual species acting as indicators, some plant and animal communities can be indicators of certain conditions. The characteristic flora of serpentine soils, which are low in calcium and high in magnesium, is a good example of a plant indicator community. In North America, for example, the serpentine soils support usually sparse, stunted narrowly endemic tree species such as *Quercus durata* and *Cupressus sargentii*. Some plant species are not tolerant to disturbance and may be used as indicators of the age of a woodland. In Britain, for example, the Nature Conservancy Council (as it was) developed a method for recording and ranking the conservation value of ancient woodlands in southern England (p. 146).

Biological indicators can be used in ecological evaluations, especially in the case of communities indicating areas of conservation interest. Indicator species are also used in environmental assessments and in the preparation of environmental sensitivity maps (p. 226).

The different kinds of biological indicators can be grouped into the following headings (Spellerberg, 1991):

1. **Sentinels**: sensitive species introduced to atypical conditions as early warning devices (canaries in coal mines);
2. **Detectors**: species occurring naturally in the area of interest which may show measurable responses to environmental change (changes in age-class, decrease in population size, behaviour etc.). Heathlands are indicators of nutrient-poor acid soils and some vascular plant species may be used as indicators of woodland type (Table 5.6);
3. **Exploiters**: species whose presence indicates the probability of disturbance or pollution (for example in aquatic habitats, the presence of many tubificid worms and or red chironomid larvae indicates polluted conditions);
4. **Accumulators**: species which accumulate (bioaccumulation) chemicals in their tissues (lichens, woodlice);
5. **Bioassay organisms**: selected organisms sometimes used as laboratory reagents to detect the presence and/or concentration of pollutants (some species of trout make suitable bioassay organisms).

2.7.2 Indices

There have been some interesting attempts to measure the overall quality of the environment. These attempts have been brought about by common concern about the effects of pollution and by a perceived environmental degradation. In the USA, there has been a rapidly growing concern since the 1970s (confirmed by many surveys) about the state of the environment. For example, the Council on Environmental Quality has published a report about public opinion on environmental issues (Council on Environmental Quality, 1980) and previous to that, the National Wildlife Federation (NWF) introduced the NWF of Environmental Quality (National Wildlife Federation, 1969). That was one of the first attempts to make a comprehensive survey of the quality of the environment. Later, a more ambitious project undertaken by the NWF was to compile a World Environmental Index (National Wildlife Federation, 1972).

Ecological assessments are undertaken to investigate the sensitivity of an area to various disturbances and the extent to which an ecosystem has been affected by disturbance or pollution. The extent to which an area has been polluted or damaged can be expressed in the form of indices: either environmental or biotic indices.

Examples of biotic indices include those which have been used to monitor water quality on the basis of indicator species and the sensitivity of species to pollution. For example in 1964, Woodiwiss described a biotic index, which is easy to calculate, based on the use of indicator species and weighted by the number of certain defined groups in relation to their sensitivity to organic pollution. The number of groups of organisms together with the presence and absence of certain indicator species is used to calculate the index. This index, now known as the Trent Biotic Index, has scores ranging from 0 (gross organic pollution) to 15 (completely free of pollution). Easy to use and cost-effective, this kind of biotic index in various modified forms has found favour with a number of water authorities in Britain (Spellerberg, 1991).

2.8 MAPPING, SURVEYING AND DATA COLLECTION

2.8.1 Classification and mapping

In as much as a standardized taxonomic classification is a prerequisite for any biological survey, so also are standardized classifications of habitats, biotopes, animal assemblages and plant associations a prerequisite for ecological evaluation. Standardized inventories and checklists vary according to the type of habitats, for example an inventory for urban open spaces (Table 2.4) is quite different from the checklist for semi-natural habitats (Table 2.5). Types of animal assemblages and plant communities can also be classified in the sense that not all kinds of woodlands are the same because they have different mixes of species. Therefore plant communities such as a woodland should be classified before being evaluated. In Britain, for example, The National Vegetation Classification (NVC) (Rodwell, 1991) has provided a phytosociological classification based on natural groupings of plants with clear ecological identities. As well as providing data from an extensive survey of vegetation, the NVC is beginning to be widely used as a standardized basis for surveys and evaluations. Another approach to classification of British vegetation has been developed by the NERC Unit of Comparative Plant Ecology (UCPE) based at Sheffield University. The method is based not on the taxonomic identity of the species present but on their functional characteristics (Grime *et al.*, 1988) and the product of that method is the Functional Interpretation of Botanical Surveys (FIBS). Examples are successional woodland dominated by species such as birch (a characteristic 'competition' species with high growth rates and with high requirements for light and nutrients) or in contrast, plants of extreme places on high mountains which are stress tolerant, slow growing and have the ability to survive long periods with limited resources.

Table 2.4 An inventory for the systematic survey or urban open spaces. From Barker (1986) with permission of G. Collins and after Sekliozotis (1980)

Urban open space use notation

Code Number	Urban open space units	
01	Heathland	1. Semi-natural Environments
02	Woodland	
03	Mossland	
04	Sand Dunes	
05	Beaches	
06	Marshland	
07	Streams and rivers	2. Water bodies
08	Canals	
09	Lakes and ponds	
10	Reservoirs	
11	Oceans	
12	Private gardens	3. Private Gardens
13	Parks	4. Amenity Open Space
14	Amenity open space – general access	
15	Amenity open space – limited access, institutional etc.	
16	Amenity open space – industrial, commercial	

17	Streets lined with trees	5. Space for Transportation
18	Streets not lined with trees	
19	Railways	
20	Motorways	
21	Open air car parks	
22	Airfields	
23	Children's playgrounds	6. Play and Recreation
24	Sportsfields and stadia	
25	Golf courses	
26	Educational playspace	
27	Industrial sports facilities	
28	Other open air playspaces	
29	Allotment gardens	7. Agriculture and Horticulture
30	Agriculture and horticulture	
31	Industrial/commercial ancillary open space (other)	8. Neglected Land
32	Rough grassland	
33	Scrubland	
34	Derelict land	
35	Mineral extraction	
36	Waste disposal sites	
37	Cleared land	
38	Cemeteries	9. Cemeteries
39	Other Open Spaces	10. Other
00	Built Environment	

Mapping biological data can be undertaken at different levels. For example it could show the distribution of plant associations, animal assemblages and habitats, or it could show the distribution of fungi on a forest floor or even microbes in the soil. Obviously we need to consider at which scale recording is most usefully undertaken: scales could range from one metre squares for plant species in a meadow to 10 metre squares in a woodland. Today there is a wealth of recording methods ranging from the various methods of remote sensing (via satellites, and aerial photography) to ground ecological surveys.

Different kinds of maps, whether they be ecological maps showing the main habitats, agricultural or geological maps showing soil types or ordinary geographical maps, can provide information for evaluations and assessments. However, prior to any evaluation of the conservation value of a region or assessment of the sensitivity of an area (the results of which may or may not be expressed on a map), it is necessary to know what is there. In addition, levels of abundance and distribution in time and space

Table 2.5 A checklist for semi-natural habitats based on Elton and Miller's (1954) classification. This has been used as the basis for standardized ecological surveys in Britain. From Bunce and Shaw (1973) with permission of R.G.H. Bunce

1 Site No.	2 Plot No.	3 Recorder	4 Date
5 Slope ° or		6 Aspect °Mag.	

A Trees – Management

7 Cop. stool	8 Singled cop.	9 Rec.cut.cop.	10 Stump hard.new
11 Stump hard.old	12 Stump con.new	13 Stump con.old	14

B Trees – Regeneration

15 Alder	16 Ash	17 Aspen	18 Beech
19 Birch	20 Hawthorn	21 Hazel	22 Holly
23 Hornbeam	24 Lime	25 Oak	26 Rowan
27 Rhododendron	28 Sweet chestnut	29 Sycamore	30 Wych elm
31 other hrwd.	32 Scots pine	33 Yew	34 Other con.

C Trees – Dead (= habitats)

35 Fallen brkn.	36 Fallen uprtd.	37 Log. v. rotten	38 Fall.bnh.>10cm
39 Hollow tree	40 Rot hole	41 Stump <10cm	42 Stump >10cm

D Trees – Epiphytes and Lianes

43 Bryo.base	44 Bryo.trunk	45 Bryo.branch	46 Lichen trunk
47 Lichen branch	48 Fern	49 Ivy	50 Macrofungi

E Habitats – Rock

51 Stone <5cm	52 Rocks 5.50cm	53 Boulders >50cm	54 Scree
55 Rock out sp.<5m	56 Cliff >5m	57 Rock ledges	58 Bryo.covd.rock
59 Gully	60 Rock piles	61 Exp.grav/ sand	62 Exp.min.soil

F Habitats – Aquatic

63 Sml.pool <1m^2	64 Pond 1–20m^2	65 Pond/ lake>20m^2	66 Strm/riv. slow
67 Strm/riv.fast	68 Aquatic veg.	69 Spring	70 Marsh/bog
71 Dtch/drain dry	72 Dtch/drain wet	73	74

G Habitats – Open

75 Gld.5–12m	76 Gld. >12m	77 Rky.knoll<12m	78 Rky.knoll>12m
79 Path <5m	80 Ride >5m	81 Track non prep	82 Track metalled

H Habitats – Human

83 Wall dry	84 Wall mortared	85 Wall ruined	86 Embankment

Table 2.5 continued

87 Soil excav.	88 Quarry/mine	89 Rubbish dom.	90 rubbish other

I Habitats – Vegetation

91 Blkthorn.thkt.	92 Hawthorn thkt.	93 Rhodo.thkt.	94 Bramble clump
95 Nettle clump	96 Rose clump	97 W.herb clump	98 Umbel.clump
99 Bracken dense	100 Moss bank	101 Fern bank	102 Grass bank
103 Leaf drift	104 Herb veg.>1m	105 Macfungi.soil	106 Macfungi.wood

J Animals (mainly signs)

107 Sheep	108 Cattle	109 Horse/pony	110 Pig
111 Red deer	112 Other deer	113 Rabbit	114 Badger
115 Fox	116 Mole	117 Squirrel	118 Anthill
119 Corpse/bones	120 Spent ctrdgs	121	122

of the plants and animals may need to be considered in that evaluation or assessment.

Ecological mapping and biotope mapping have been especially popular in some European countries such as the Netherlands and in Germany. In the Netherlands, early initiatives in ecological and environmental mapping (or 'millieukartering' as it was called commenced as long ago as 1968 and today landscape ecological mapping and evaluation has become an essential part of the planning process. Ecological mapping projects in the Netherlands were attempted on the national grid system (25 km²) so that they would be compatible with mapping of the flora and birds. The ecotope concept has been a central theme for mapping in the Netherlands, that is, vegetation-related data are collected on the basis of an ecotopic landscape pattern. Ecotopes are matched ecotopic complexes or large ecological landscape units. They are provisionally differentiated on the grounds of soil characteristics and landscape observations and by means of their identification through vegetation complexities and plant associations (Maarel and Stumpel, 1974).

Detailed vegetation surveys and mapping of flora and fauna in the Netherlands is now undertaken not at a national level but by each of the provinces (Han Runhaar personal communication). Although there is some coordination, every province uses its own type of mapping method. For example, in northern Holland (e.g. Utrecht and Zeeland), the basic method is based on lists of species found in certain landscape elements within each square kilometre. This results in separate checklists for different areas such as grasslands, marshes, road verges and forests. In southern Holland the vegetation is described on the basis of relevees; about 20 relevees for each square kilometre. The vegetation in each km² is described by means of a reference to these relevees that are chosen in such a way as to be representative of the variation in the km². The surveying and mapping at the province level has produced only a fragmented and incomplete account and consequently there has been a need for coordination. Such coordination is now undertaken at the Centre for Environmental Sciences at Leiden and the aim has been to try and convert data to one standardized classification.

In Germany, two methods of biotope mapping (selective and comprehensive) have been employed as a basis for evaluation of city environments (Sukopp and Weiler, 1988). The former refers to the fact that only biotopes worthy of protection are examined (namely 'especially valuable'; Wittig and Schreiber's method, p. 150). The latter is not confined to selective biotopes and in contrast the aim is to establish an inventory of all habitats. Non-evaluative in the first stage, the comprehensive biotope mapping as used in some German cities can provide a sound basis for ecological evaluation and for ecological monitoring. Based on units of a maximum of 4 ha, the comprehensive biotope mapping includes inventories of wild plant species, vegetation types and inventories of breeding birds (all recorded in relation to defined land use types). The last of these has been suggested on the grounds that breeding birds are good indicators of individual biotopes (Sukopp and Weiler, 1988). The many values and benefits of wildlife in cities (amenity, ecological, environmental) have been used to justify such detailed surveys (Sukopp *et al.*, 1990). The costs of comprehensive biotope mapping have been criticized but it seems that as yet no-one has attempted to undertake a cost–benefit analysis and quantify the benefits (in monetary terms) arising from such mapping.

2.8.2 Baseline data

The information for ecological evaluations and assessments can come from two sources; surveys conducted for that purpose or records and publications already in existence. There would appear to be a growing demand for the latter, particularly in a form which will help to interpret any changes that may be detected at a later date. That is, initial surveys or baseline information can be an important aspect of ecological assessments. Over the last few decades there has been a growing number of organizations which can not only supply biological and ecological information but can do so in an ever-increasing professional manner.

In Britain there has been a wealth of studies in natural history over the last few centuries. However, there are surprisingly few areas of the country for which there are recent and comparatively detailed biological and ecological surveys which would provide baseline data for ecological evaluations. There are some noteable exceptions in the form of county-wide surveys. For example, the County of Warwick was the subject of a biological and ecological survey for three years from the summer of 1975. One of the main aims of that ambitious project was to try to quantify the extent and state of every habitat in the county and so provide a baseline for measurement of future change (a lesson which could well be noted by other counties). At the same time it was hoped that the survey would help to identify the relative importance of the more important habitats and draw attention to the value of the wildlife. The amount of data that were collected is a tribute to the energy and enthusiasm of not only the organizers of the survey but to a large assembly of volunteers. The form of data in the subsequent publication (Copson and Mitchell, 1980) is a helpful guide as to how data can be collected and presented (Box 2.1).

Over ten years later, the survey of the Warwickshire countryside and especially the original data have proved useful and because there has been no similar survey work in the county, the 1975/78 survey provides primary and baseline data for biological recording and for planning (P. Copson, personal communication). Monitoring change in the countryside requires good baseline data such as these and indeed the Warwickshire survey has been particularly useful in monitoring hedgerow losses from a map base in 1956 through to the survey in 1976 and to the present. Although no new survey is planned, biological data are being updated by site as part of the work undertaken for the Warwickshire Biological Records Centre. This kind of data gathering exercise and more especially the updating of the information does require a lot of time and effort. Given the facilities and the training, this kind of survey work could be continued in a cost-effective manner using information technology and Geographical Information Systems (GIS) which facilitate the storage, spatial analysis and communication of map data.

2.8.3 Factors affecting sampling

All recording and sampling methods have their limitations and are all biased towards certain taxonomic groups. Two useful concepts relevant to measures of bias in sampling methods were recognized by Disney (1986): collecting success and collecting efficiency. The first refers to the number of species which are collected per unit of sampling effort. The second concept refers to the proportion of the total number of species present that are actually caught. It is important therefore to appreciate how various recording and sampling methods may affect the outcome of an ecological evaluation. For example, pitfall traps (basically a plastic beaker or cup buried in the ground with the lip of the beaker at ground level) have long been used in field ecology for sampling ground dwelling invertebrates. It is frequently assumed that, provided the traps are placed correctly in the ground, they will catch a representative sample of the invertebrate fauna. However, the catch in a pitfall trap is a function of both numbers and

Box 2.1 An example of the data collected for the Warwickshire Countryside Ecological Evaluation (Copson and Mitchell, 1980). This is not an ecological evaluation in the sense used in this book but Copson and Mitchell have shown how environmental data can be assembled and presented in a useful manner.

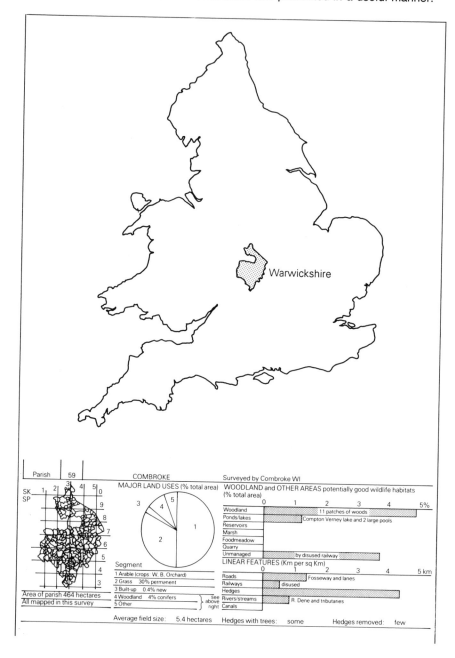

Box 2.1 continued

List of Special Ecological Sites: Notes on the Ecology:

9/35 Oxhouse Nature Reserve	The valleys of the Dene and Combrook are incised into the Lower Lias clay plateau. The
10/35 Compton Verney Lake & Woods	plateau is given over to cereal growing, smallish fields separated by elm-lined hedges whose
20/35 River Dene	future must be limited. Two coverts and two quite
35/35 Disused Railway line	large ponds are attractive for wildlife within this arable area. The two river valleys are very rich areas indeed for flora and fauna. Compton Verney Lake is a refuge for wildfowl and stocked for private fishing. The surrounding woods contain some fine trees and are also good for woodland birds although quite substantial areas have now been replaced by commercial conifers. Below the village are some flower-rich unspoiled meadows. The Dene valley has the added bonus for wildlife of a parallel disused railway line now parrtly managed as a private nature reserve. The cuttings into White Lias strata produce a rich calcecole vegetation and associated fauna. Boulder clay caps higher land to the east.

activity and the efficiency with which the trap captures and retains individuals: this varies from species to species (Halsall and Wratten, 1988). Therefore, the catch does not provide reliable estimates of the relative numbers of the different species in the habitat. It is important therefore to recognize and to be able to quantify the limitations and biases of sampling and recording equipment.

In Chapter 5, evaluation methods for heathlands are mentioned and the importance of noting the timing of sampling is emphasized, especially in connection with one worker's use of invertebrates (p. 24). In that instance the invertebrates were sampled with a D-vac (a large hand-operated sampling device with a powerful suction mechanism) and different catches were obtained at different times of the year and even at different times of the day. As well as temporal and spatial considerations, the location of sampling equipment in the habitat can have important implications. This was illustrated by Usher (1990) who investigated the use of water traps for sampling arthropod communities of heather moorland. Water traps are simply dishes or small containers, containing water and detergent which when laid out on the ground will collect flying insects and other arthropods. The influence of size and colour of the dishes and other factors affecting the catch are discussed in Southwood (1978). Usher assessed the use of water traps with respect to different areas of moorland (all managed in a traditional manner), the development or growth phase of the *Calluna*, and the height of the water traps above the vegetation. He found that the height was extremely important in influencing the number of individuals caught and the diversity of Diptera (at the family level). The implications are clear: as well as standardizing trap colour and size it is also important to standardize the height of the trap above the vegetation.

2.9 CONCLUSIONS

Two requirements emerge from this brief and selective account of ecology; one is the need for a standardized methodology and the other is the need to state the variation in the results. Clearly the value of various sites can only be compared if the same data are collected in the same way, right down to the detailed methods of sampling and methods of identification of species and communities. Without standardization, there can be no justification for comparisons. When making comparisons, based on ecological data, some measure of natural variation should be available. For example if species richness is used as a criterion and if areas are to be compared then obviously the variation in species richness of various groups at the different sites should be quantified. Assume for example that the two sites A and B (when examined on one occasion) have a species richness for birds of 18 and 23 respectively. On the surface, site B would seem to have a higher species richness and possibly greater conservation interest. More detailed sampling, however, might show that variation in species richness from year to year varies at site A from 17 to 26 and 19 to 24 at site B.

Ecological evaluations and assessments could be said to be interdisciplinary in their application because they are of interest to a range of professions including politicians, planners, environmental economists, ecologists and conservationists. It is of great importance therefore that the evaluation methods and results can be understood by this range of people. However, many of the methods described in this book are based on ecological science and obviously there will be people using ecological evaluations who do not have a background in ecology. I have tried to address this problem by including this chapter but I feel that a better solution is to foster more interdisciplinary training such as that offered by environmental sciences and environmental studies courses.

REFERENCES

Anderson, J.A. (1991) A conceptual framework for evaluating and quantifying naturalness. *Conservation Biology*, 5, 347–52.

Barker, G.M.A. (1986) Biological survey and evaluation in urban areas: methods and application to strategic planning. Proceedings of a conference, Polytechnic of North London. NCC, Peterborough.

Bunce, R.G.H. and Shaw, M.W. (1973) A standardized procedure for ecological survey. *Journal of Environmental Management*, **1**, 239–58.

Copson, P. and Mitchell, J. (1980) *The Warwickshire Countryside an Ecological Evaluation*. Warwickshire Museum Service, Warwick.

Council on Environmental Quality (1980) Public Opinion on Environmental Issues, Washington DC.

Disney, R.H.L. (1986) Assessments using invertebrates: posing the problem, in *Wildlife Conservation Evaluation* (ed. M.B. Usher) Chapman & Hall, London, pp. 271–93.

Elton, C.S. and Miller, R.S. (1954) The ecological survey of animal communities: with a practical system of classifying habitats by structural characters. *Journal of Ecology*, **42**, 460–96.

Fisher, R.A., Corbet, A.S. and Williams, C.B. (1943) The relation between the number of species and the number of individuals in a random sample of an animal population. *Journal of Animal Ecology*, **12**, 42–58.

Fitter, R. and Fitter, M. (1987) The road to extinction. Problems of categorizing the status of taxa threatened with extinction. IUCN, Gland.

Grime, J.P., Hodgson, J.G. and Hunt, R. (1988) *Comparative Plant Ecology: a Functional Approach to Common British Species*. Unwin Hyman, London.

Halsall, N.B. and Wratten, S.D. (1988) The efficiency of pitfall trapping for polyphagous predatory Carabidae. *Ecological Entomology*, **13**, 293–9.

Hawksworth, D.L. and Rose, F. (1976) *Lichens as Pollution Monitors*. Edward Arnold, London.

Helliwell, D.R. (1983) The conservation value of areas of different size: Worcestershire ponds. *Journal of Environmental Management*, **17**, 179–84.

Kershaw, K.A. and Looney, J.W.H. (1985) *Quantitative and Dynamic Plant Ecology*, 3rd edn, Edward Arnold, London.

Lewis, T. and Taylor, L.R. (1964) Diurnal periodicity of flight by insects. *Transactions of the Royal Entomological Society, London*, **116**, 393–476.

Luff, M.L., Eyre, M.D. and Rushton, S.P. (1989) Classification and ordination of habitats of ground beetles (Coleoptera, Carabidae) in north-east England. *Journal of Biogeography*, **16**, 121–30.

Maarel, E. van der and Stumpel, A.H. (1974) Landschaftsokologische kartiening und bewertung in den Niederlanden. *Verh. Ges. Okol. Erlangern*, 231–40.

Magurran, A.E. (1988) *Ecological Diversity and its Measurement*. Croom Helm, London.

Margalef, R. (1951) Diversidad de especies en las comunidades naturales. *Publnes. inst. Biol. apl., Barcelona*, **6**, 59–72.

Margules, C. and Usher, M.B. (1981) Criteria used in assessing wildlife conservation potential: a review. *Biological Conservation*, **21**, 79–109.

Menhinick, E.F. (1964) A comparison of some species–individuals diversity indices applied to samples of field insects. *Ecology*, **45**, 859–61.

National Wildlife Federation (1969) NWF Index of Environmental Quality. *National Wildlife*, **7**, 2–13.

National Wildlife Federation (1972) World Environmental Quality Index. *International Wildlife*, **2**, 21–36.

O'Connor, K.F., Overmars, F.B. and Ralston, M.M. (1990) *Land Evaluation for Nature Conservation*. Department of Conservation, Wellington.

Perring, F.H. and Walters, S.M. (1962) *Atlas of the British Flora*. BSBI/Nelson, London.

Pielou, E.C. (1966) The measurement of diversity in different types of biological collections. *Journal of Theoretical Ecology*, **13**, 131–44.

Rabinowitz, D. (1981) Seven forms of rarity, in *The Biological Aspects of Rare Plant Conservation* (ed. H. Synge), Wiley, Chichester, pp. 205–17.

Rodwell, J.S. (1991) *British Plant Communities*. Vol 1. Cambridge University Press, Cambridge.

Savage, A.A. (1982) Use of water boatmen (Corixidae) in the classification of lakes. *Biological Conservation*, **23**, 55–70.

Sekliozotis, S. (1980) A survey of urban open spaces using colour infra-red aerial photographs. PhD Thesis, Aston University, UK.

Simpson, E.H. (1949) Measurement of diversity. *Nature*, **163**, 688.

Smith, P.G.R. and Theberge, J.B. (1986) A review of criteria for evaluating natural areas. *Environmental Management*, **10**, 715–34.

Sneath, P.H.A. and Sokal, R.R. (1973) *Numerical Taxonomy. The Principles and Practice of Numerical Classification*. Freeman, San Francisco.

Southwood, T.R.E. (1966) *Ecological Methods, with Particular Reference to the Study of Insect Populations*. 2nd edn, Chapman & Hall, London.

Specht, R.L., Roe, E.M. and Boughton, V.H. (1974) Conservation of major plant communities in Australia and Papua New Guinea. *Australian Journal of Botany*, Supplement 7.

Spellerberg, I.F. (1991) *Monitoring Ecological Change*. Cambridge University Press, Cambridge.

Sukopp, H. and Weiler, S. (1988) Biotope mapping and nature conservation strategies in urban areas of the Federal Republic of Germany. *Landscape and Urban Planning*, **15**, 39–58.

Sukopp, H. (1990) *Urban Ecology*, SPB Academic Publishing, The Hague.

UNESCO (1974) Criteria and guidelines for the choice and establishment of biosphere reserves. *MAB Report Series* No. 22. UNESCO, Paris.

Usher, M.B. (1990) Assessment of conservation values: the use of water traps to assess the arthropod communities of heather moorland. *Biological Conservation*, **53**, 191–8.

Webb, N. (1986) *Heathlands*. Collins, London.

Woodiwiss, F.S. (1964) The biological system of stream classification used by the Trent River Board. *Chemistry and Industry*, 443–7.

Wright, D.H. (1990) Human impacts on energy flow through natural ecosystems and implications for species endangerment. *Ambio*, **19**, 189–94.

3

Methods of valuing nature and the environment

3.1 INTRODUCTION

This book is concerned mainly with evaluating biodiversity for conservation. It would seem useful to put this aim into perspective by considering briefly some of the methods which have been proposed for valuing species and the environment. The direct or indirect contributions of the environment and biodiversity to a country's economy have long been of interest. For example, there have been many attempts to quantify the direct contribution to a country's economy from wild species, wild harvested resources and indirect benefits from ecosystems. The direct value of wildlife can be of different types (Table 1.5), that is the value of game animals or firewood is a consumptive use value and the value of commercially harvested wildlife such as fish and medicinal plants is the productive use value. Watershed forests, wind breaks and marshes acting as flood barriers all have indirect values which are non-consumptive.

Valuing the environment and biodiversity is both an exciting and difficult interdisciplinary subject to which many environmental economists, ecologists and conservationists have contributed. However, only more recently have studies been carried out in this area (see, for example, Johansson, 1987; Costanza et al., 1989; Pearce and Markandya, 1989; Pearce et al., 1989, 1991; Winpenny, 1991; de Groot, in press). This chapter considers some methods of giving monetary figures (e.g. market price, replacement cost, willingness to pay etc.) to wildlife and natural areas and discusses the limitations or advantages of various methods. Some of these methods are concerned mainly with valuation in terms of direct economic benefits and others are directed at identifying and attaching a monetary value to non-market benefits of wildlife. Although not all these methods would be taken seriously by environmental economists, they have been chosen because they represent a range of ideas from both ecologists and economists.

3.2 NOTIONAL MONETARY VALUES

While researching priorities and values in nature conservation, Helliwell (1973) proposed some methods for the comparison of one wildlife resource with another together with a suggestion of attaching notional monetary values to those resources. A basic method of valuing individual wild species was included in his proposed methods (Table 3.1). In that method, the value of oak (*Quercus robur*) as a species in Britain would be calculated as follows:

32 (many associated species) × 16 (very common) × 4 (very conspicuous) × 4 (some material value) = 8192.

Helliwell suggested that the score should then be multiplied by the arbitrary figure of £10 000 to give a shadow price and went on to suggest that the total value for all species of plants and animals in Britain should be approximately £5000 to £6000 million.

Various weightings may be included in these calculations. For example Helliwell (1985) suggested that abundance scores could be weighted so that 'warm blooded' animals rate twice as much as 'cold-blooded' animals and predatory species be given five times the rating of herbivorous animals. The logic behind this weighting lies in the fact that the term 'abundance' is used by Helliwell as a composite term covering both numbers and size, rather than strict numbers alone. That is, a butterfly with 1000 individuals could be classed as very rare but a species of deer represented by 1000 individuals could be considered as locally common.

There are two questions which arise from this: (1) why use these notional monetary values and (2) is the method meaningful? We could argue that monetary evaluation should be used because the incentives for new developments and changes in land use are, on occasions, financial. There seems to be a good case for arguing that we should give careful consideration to the use of monetary evaluation of wildlife at times when alternative uses of land are justified in monetary terms. As far as the method is concerned, it is easy to be critical of this method of evaluation because it appears to be completely arbitrary in every stage. Although Helliwell did admit to the use of crude categories we might ask is it just a set of methods or game which some conservationists have to play because others play that game? The fundamental point is that society when faced with a choice, say between building houses or conserving a heathland, has to decide which of these uses will yield the greatest net benefit now and in the future. That is no easy task and one which requires the development of new principles by both economists and ecologists.

Table 3.1 Valuation of individual wild species. After Helliwell (1973), with permission of the author

	Scores					
Factor	1	2	4	8	16	32
Associated species	Very Few Brent Goose (*Branta bernicla*) Snowdon lily (*Lloydia serotina*)	Few Primrose (*Primula vulgaris*) Golden eagle (*Aquila chrysaetos*) Knott (*Calidris canutus*) Sycamore (*Acer pseudoplatanus*)	Some Cock's foot grass (*Dactylis glomerata*) Green woodpecker (*Picus viridis*) Stoat (*Mustela erminea*)	Several Blue tit (*Parus caeruleus*) Robin (*Erithacus rubecula*) Stinging nettle (*Urtica dioica*) Badger (*Meles meles*)	Many Rowan (*Sorbus aucuparia*) Field maple (*Acer campestre*)	Very many Oak (*Quercus robur*) Birch (*Betula verucosa*)
Abundance in Britain	Very Rare Red kite (*Milvus milvus*) Snowdon lily (*Lloydia serotina*)	Rare Pine marten (*Martes martes*) Bird's eye primrose (*Primula farinosa*)	Uncommon Sparrow hawk (*Accipiter nisus*) Cloudberry (*Rubus chameamorus*)	Common Badger (*Meles meles*) Common reed (*Phragmites communis*)	Very common Chaffinch (*Fringilla coelebs*) Wren (*Troglodytes troglodytes*) Stinging nettle (*Urtica dioica*)	

Table 3.1 continued

Conspicuousness	Inconspicuous	Reasonably noticeable	Very conspicuous or attractive
	Lichen (*Peltigera canina*) Annual poa (*Poa annua*)	House sparrow (*Passer domesticus*) Cow parsnip (*Heracleum sphondylium*)	Deer Bird's eye primrose (*Primula farinosa*) Golden eagle (*Aquila chrysaetos*)
Material value	None House sparrow (*Passer domesticus*) Cow parsnip (*Heracleum sphondylium*)	Little Rubus spp. Vaccinium spp. Coypu (*Myocastor coypus*) Common reed (*Phragmites communis*) Field maple (*Acer campestre*)	Some Trout (*Salmo sp.*) Red deer (*Cervus elaphus*) Grouse (*Lagopus spp.*) Scots pine (*Pinus sylvestris*)

3.3 VALUING ECONOMIC BENEFITS

Estimates of consumptive and productive use values of wildlife are comparatively straightforward. For example the value of commercial fish stocks and other marine life is well documented in many countries. Similarly the revenue generated by angling and by shooting wild game birds and mammals has long been of interest not only to sporting organizations but also to the wild game market. The value of forests and its wild game animals has been of particular interest to some European countries because of the economic importance of those animals. There have been several estimates of the value of wild game animals such as various species of deer, wild boar (*Sus scrofa*), hares and rabbits, pheasants (*Phasianus colchicus*) and other game animals. For example Nusslein (1974) produced figures for the production (number of animals killed as game) and value of free-roaming game animals subject to hunting legislation (in German these are described as 'freilbende Tiere, die der Jagdgesetzgebung der einzelnen Lander unterliegen'). The values excluded revenue generated by licences and taxes. Using information collected for 1969–1972 from several European countries, Nusslein was able to give estimates of the wild animal production and their market value in DM. For that period of time the animal production came to about 79 million kilograms with a market value of DM 328 million (Table 3.2). Nusslein goes on to consider the value of wild game animals

Table 3.2 Wildlife production and value. Data for three years (1969–72) for Europe. From Nusslein (1974)

Country	Area (millions ha)	Wildlife production (millions kg)	Value (millions DM)
Germany	35.6	22.0	96.7
Denmark	4.2	3.7	14.6
Finland	33.7	1.6	5.2
Jugoslavia	25.6	3.7	11.3
Luxemburg	0.3	0.1	0.5
Norway	32.4	2.5	11.6
Austria	8.4	6.3	27.0
Poland	31.2	6.9	29.6
Romania	23.8	2.9	10.7
Sweden	45.0	8.0	35.3
Switzerland	4.1	0.7	3.4
Spain	50.4	9.0	34.8
Czechoslovakia	12.8	8.2	32.0
Hungary	9.0	3.0	15.0
Total for 14 Countries	316.5	78.6	327.7

in forests only (DM 148 million) but he also made the observation that the 'non-economic' value of game and wildlife should not be overlooked, that is, there are values which arise from economic, sociopolitical and ethical factors.

The importance of biodiversity for agriculture and for medicine is often underestimated. Plant breeding in agriculture has often turned to wild genetic material and in North America for example there has been increasing dependence on genetic material from wild relatives of maize and wheat (see p. 18). In 1980 the value of increased productivity attributable to contributions from such wild material was estimated to be around a billion dollars. In the 1980s, about half of all prescribed medicines had their origins in wild plants and animals and the commercial value of those pharmaceuticals in the United States was estimated at the time to be around 14 billion dollars. On a world basis the figure is put at about 40 billion dollars (WCED, 1987).

Perhaps not surprisingly the methods of valuing species have attracted some attention from economists. For example, Brown and Goldstein (1984) have studied the minimum required information if society is going to rationally conserve some species and not others. In describing a model for valuing species they draw attention to the fact that species are not independent so that there are opportunity costs, both positive and negative. These are created by relationships between species in ecosystems and removal of some species. Such values are difficult to measure and it is an area of research that will receive increasing attention from both ecologists and economists. A particular challenge is to find the analytical and empirical framework enabling calculations of probable economic values on specific species (Brown and Goldstein, 1984; Aylward, 1991).

Methods for valuing natural areas such as wetlands have been well researched (Aylward and Barbier, 1991; Winpenny, 1991). Certainly, the economic value of some wetlands seems to be very clear. For example, the mangrove swamps in Bangladesh not only provide a source of trees and fish worth in excess of 10 million US dollars per year but also protect the interior from tidal surges (Paul *et al.*, 1989). It is not impossible to determine the value of those ecosystems which function as buffers against storms and floods. The wetland areas of Boston, MA (USA) which act as buffers against floods have realized annual cost savings of 17 million dollars according to the US Army Corps of Engineers (Hair, 1988, cited in McNeely, 1988).

3.4 AMENITY EVALUATION

In relation to planning undertaken by local authorities and conservation organizations, there is sometimes a need to assess the relative value of natural areas (sometimes called 'green areas') in relation to other natural

Box 3.1 The Helliwell System for valuation of amenity woodlands. Reproduced from the Arboricultural Association publication *Amenity Valuation of Trees and Woodlands*, 1990 with kind permission of the Association and the author.

Evaluation method for woodlands

Six standard factors are identified for each woodland area, plus any special factors such as historical association, accessibility, or the screening of unwanted views. For each of these factors the woodland is given a score of up to four points, and the scores for all the factors are then multiplied together to give an assessment of the amenity value of the woodland.

	Points			
Factor	1	2	3	4
i. Size of woodland	very small	small	medium	large
ii. Position in landscape	secluded	average	prominent	very prominent
iii. Viewing population	few	average	many	very many
iv. Presence of other trees and woodland	area more than 25% wooded	area 5–25% woooded	area 1–5% wooded	area less than 1% wooded
v. Composition and structure of the woodland	dense plantation or blatantly derelict woodland	even-aged pole-stage crops with mixed species	semi-mature or uneven-aged woodland with fairly large trees	mature or uneven-aged woodland with large trees
vi. Compatibility in the landscape	just acceptable	acceptable	good	excellent
vii. Special factors	none	one	two	three

Explanatory notes on these factors are given below.

If a woodland is less than 0.1 hectare it should be evaluated as a group of individual trees (or as one very large 'tree') and not as a woodland, using the Evaluation Method for Trees.

Woodlands which (due to their position, shape, or other reason) are unacceptable in the landscape should score zero under factor 6. This is not to say that a poorly-shaped woodland could not be an amenity if it were modified in some way. It should be stressed that this method should be applied sensibly, with due regard to such points. It would, for example, be possible to assess the amenity value of a woodland in 20 years' time, on the assumption that an additional area is planted now, and to use the assessment to arrive at a decision on the giving of grant aid for such planting.

Box 3.1 continued

As with the evaluation method for trees, some factors can be assessed with a large degree of objectivity. For example, the abundance of woodlands in the area, and the condition of the woodland will usually be fairly clear. With some factors, however, there is bound to be an element of subjectivity, and there may be a moderate amount of disagreement between different people's assessments.

Again, this method of assessment only takes account of the amenity value of a woodland, and does not assess any costs which may be incurred in managing the woodland. Such costs are usually ascertainable, and will also need to be considered before management decisions are taken.

Explanatory notes

(a) The size of the woodland is taken as the area which is visible from the vantage points being considered. On steep hillsides the full area of the woodland will be taken. On flat land, where only the outer wedge of the woodland is visible, and there is no access into the wood, a band 50m wide at the edge of the wood should be taken.

a 'small' woodland is between 0.1 and 0.5 hectare.

a 'large' woodland is between 2 and 10 hectares.

a 'very large' woodland is between 10 and 40 hectares.

Woods in excess of 40 hectares should be evaluated in more than one unit; and where different parts of a wood are of different quality or visibility these should also be evaluated separately.

If greater degree of refinement is sought, the relevant points score can be taken from the following scale.

size of woodland (ha)

0.1	0.25	0.5	1	2	4	8	15	20	30	40
points										
0.75	1	1.5	2	2.5	3	3.5	3.8	4	4.3	4.5

and the values for other sizes interpolated.

(b) Position in landscape

A 'secluded' woodland is one which is visible from a restricted area of not more than about one square kilometre.

An 'average' woodland will be visible, but not particularly prominent, over an area of one to four square kilometres.

A 'prominent' woodland will be **readily** visible over an area of at least two square kilometres; and a 'very prominent' woodland over an area of more than four square kilometres.

(c) Calculations of 'viewing population' are unlikely to be exact. A sensible procedure may be to take 2% of the resident population within sight

Box 3.1 continued

of the woodland; plus one person per vehicle present on roads at any one time during normal traffic conditions; plus the total numbers of people on foot at any one time. Thus, a woodland on the edge of a town, within sight of about 300 houses, containing about 1000 people; visible from about 2km of motorway; and from several footpaths and play areas may have a 'viewing population' of $20 + 50 + 30 = 100$.

The following figures are suggested for the various categories:

'few'	=	0.2 viewing population
'average'	=	2–20 viewing population
'many'	=	20–100 viewing population
'very many'	=	100 + viewing population

(d) Presence of other trees and woodland
This factor is included to allow for the decreasing amenity importance of any individual area of woodland as the amount of other trees and woodland in the vicinity increases. (Maximum overall amenity values for a parish or district are often obtained with 50 to 60% woodland cover.)
The area which should be assessed for tree and woodland cover should normally be four square kilometres.

(The woodland being assessed should not be included in the percentage figure.)

(e) Composition and structure of the woodland
This is intended to be a purely visual assessment, giving young or monotonous plantations a low score and diverse or mature woodland a high score. A pure even-aged spruce plantation is likely to be classed as monotonous and uninteresting at almost any stage, unless it is allowed to grow on past the normal age for commercial felling, and will score no more than 1; but a mature pine plantation containing a mixture of birch and with some open gaps may be quite attractive and score 3 or 4. However, it should be emphasized that there is no reason why, **in the right place and with appropriate management**, a plantation which is predominantly composed of spruce, or any other species, should not score highly under this factor. It should also be noted that an uneven-aged woodland is more likely to **retain** a high amenity value than an even-aged woodland, which is likely to be clear-felled.

(f) Compatibility in the landscape
As noted above, woodlands which fit very badly into the landscape should receive a zero score. Examples of such woodlands might be rectangular shelterbelts running contrary to the lie of the land in upland areas, and 'pyjama stripe' mixtures on steep hillsides.

Woodlands which are 'just acceptable' might include coniferous plantations in a predominantly broadleaved landscape, or woodlands of any type which

Box 3.1 continued

> are of a rigid or awkward shape. In most cases it would be easy to envisage ways in which a modest change in species, management, or shape of the woodland would improve its appearance.
>
> 'Good' and 'excellent' ratings might be given to woods which fit so well into the landscape that it is diffcult to imagine the landscape without them.
>
> **(h) Special factors**
> These will include easily accessible woods with public rights of way or open access; well-known landscape features; woods which help to screen unsightly views; and woods with notable displays of wild flowers.

areas. In other words, from time to time and in relation to planning new developments, the areas highly valued for their amenity will need to be identified so that decisions can be made about their management.

One method used to determine the value of amenity woodlands has been developed in Britain. This method, which is published by the Arboricultural Association and is endorsed by the Tree Council, was first devised as long ago as 1967 by Rodney Helliwell. The objectives are twofold. First, to provide a disciplined format for the assessment of the amenity value of a woodland and second, to give planners and managers a basis for including precise and consistent amenity values in the calculations.

Six standard factors are identified for each woodland and for each of these factors there is a score of up to four points (Box 3.1). The scores for all the factors are then multiplied together to give an assessment of the amenity value of the woodland. The factors are as follows: size of woodland, position in the landscape, viewing population, presence of other trees and woodlands, composition and structure of the woodland, and compatibility in the landscape. Special factors such as historical associations and accessibility are also included.

Market values of economically important species can easily be calculated, but what of those species or communities which are at the time of no direct economic importance? The method described above for determining the amenity value of a woodland can be extended to a monetary valuation. As at 1990, for example, a figure of £25 was recommended as a conversion factor. The total amenity score would simply be multiplied by 25 to give a monetary value.

Environmental economists have and continue to take a particular interest in methods for evaluating environmental amenity and a whole range of other environmental benefits (see for example Johansson, 1987; Pearce and Markandya, 1989). Probably the most useful technique of the many available is that known as the Contingent Valuation Method (CVM), for which there are a number of techniques (Winpenny, 1991). Certainly the CVM has had increasingly common use in the measurement of environmental benefits and costs. It aims to elicit valuations or 'Bids' and basically asks people what they are willing to pay for a benefit and what they would be willing to receive by way of compensation to tolerate a cost. This information can be obtained by way of questionnaires or from experimental techniques in which respondents are exposed to various artificial conditions. Pearce and Markandya (1989) give an interesting example in which a Contingent Valuation study indicated a West Berlin evaluation of clean air of DM 4.6 billion (about 7% of GDP). That example from research (Schulz, 1985) was undertaken by way of a willingness to pay (WTP) survey in which the WTP questions used two hypothetical payment methods: one in which polluters bore all the costs and one in which the respondent would pay for abatement of air pollution. Perhaps not surprisingly, younger people placed significantly more value on clean air. One other interesting example is described by Jacobs (1991) with regard to wildlife. It was found that respondents to a questionnaire in the USA who had no intention of ever seeing grizzly bears (*Ursus arctos*) in the wild were, however, willing to pay an average of $24 a year to ensure their preservation. This existence value was greater than the use value generated by bear hunters ($21.50).

Contingent valuations by way of questionnaires can only be as good as the questionnaire methodology and in addition there are other factors which could bias the level of willingness to pay. The method of payment, for example, could influence that level. For example, the continued designation of a nature reserve could be paid for by an entrance fee, by contributions to a conservation society, by an increase to tax and so on. Respondents may feel that one method is more appropriate than another method and give their valuations accordingly (Jacobs, 1991).

3.5 ENERGY-BASED EVALUATION

The flow of energy through the environment and through ecosystems makes possible the circulation of money and relationships between energy and money are basic to the understanding of economic systems. This theme was taken up by Odum and Odum (1976) who explored various relationships between energy and humankind and who also looked at flows of energy. It has long been known that money is involved only in some stages of energy flow but that we are dependent on the flow of energy through entire ecosystems. That being the case, it is perhaps not surprising that there have been attempts to develop energy value theories. The basic assumption in energy value theories is that natural areas are a necessary and essential component of our whole environment and that sets of costs in ecosystems can be determined by appropriate allocation of the net input to the ecosystem (Hannon *et al.*, 1986; Costanza *et al.*, 1989). In one approach to energy-based evaluation, Nijkamp (1977) described the exchange of energy and money as being at the basis of the approach. The ratio of the Gross

National Product (GNP) to National Energy Consumption (NEC) was used to equate energy with money. Thus the unit value of energy (UVE) was derived by Nijkamp as follows:

$$UVE = GNP/NEC$$

The amount of primary production (see glossary Appendix C) in an ecosystem is a measure of energy flow (EF) within the ecosystem and this amount of energy can be used to evaluate, in monetary terms (but not in economic terms), any part of a natural area. (It is relatively easy to measure primary production and standard techniques for this has been developed by ecologists.) The total energy value (TEV) of an ecosystem is derived as follows:

$$TEV = EF \times UVE$$

This represents the annual value of an ecosystem and therefore the income and capital value of a natural area such as a wetland can be calculated and used in planning. An application of this approach has been reported by Nijkamp (1976, 1977) with reference to some wetlands in the Netherlands. In the northeast of the Netherlands, a plan had emerged to improve the drainage of some agricultural land by constructing a canal either through a wetland or around the wetland. The former, in simple construction cost terms, would have been the cheaper alternative.

Construction of the canal through the wetland would have required about 600 ha to be destroyed. Nijkamp and Verhage (1976) described how the energy-based value of that area of wetland was calculated. In 1974, the gross national energy consumption in the Netherlands was about 705 $\times$ 10^{12} kcal. The conversion ratio between production and energy was then about 1: (4.1×10^3). In 1974 the annual value of 1 hectare of the estuary was calculated as being $10^8 / (4.1 \times 10^3) = 24.4 \ 10^3$ Dfl.

Apparently the Dutch Government decided not to proceed with the canal through the wetland, despite the immediate direct costs of constructing a canal around the wetland. In this instance, therefore, the energy-based value of a natural area was used in a planning decision despite the fact that the energy involved is of no direct or indirect value to the inhabitants of Holland.

3.6 REPLACEMENT VALUE: SHADOW-PROJECT SCHEMES

In circumstances where a proposed development will eventually damage or destroy an area of conservation importance, one potential solution would seem to be the establishment of the habitats elsewhere (a form of creative conservation). This is commonly done in the USA to mitigate environmental impacts. The value of the original area can therefore be based on the costs involved in its development and substitution. The shadow-project scheme involves a detailed estimate of the costs required to construct a similar environment: fragile or rare habitats would predictably cost more than common or resilient habitats. Not surprisingly, shadow-project schemes may be costly to research and to pursue because of the high construction costs, the costs of long-term monitoring and compensation for the time during which the natural area is lost to the public. Nevertheless, shadow-project policies have been employed in the Wadden Sea area of the Netherlands in connection with the construction of Europoort (Nijkamp, 1976).

The idea is to value the natural area on the basis of how much it would cost to replace but not necessarily to provide the same kind of habitats. However, if there were any intention to actually establish new habitats then the realities of such an intention should be assessed. Determination of the value of an area of conservation interest by calculating the costs involved in reconstructing that area elsewhere is perhaps based on a naive understanding of ecology. Superficially similar habitats can be created but it is impractical to recreate a habitat which includes precisely the same species composition, population structures and distribution of all the taxonomic groups. On the other hand, it has been possible to relocate populations of certain species and at the same time modify an existing habitat to accommodate the relocated species. In other words, at the species level it is relatively easy to estimate costs but at the ecosystem level it is an exercise which would require sophisticated modelling techniques together with very good costing methods.

3.7 CONCLUSIONS

The Helliwell method for calculating notional monetary values has four basis stages: the identification of 'important' ecological or biological variables, allocation of arbitrarily chosen numerical scores to the variables, combining the scores and multiplying the scores by a monetary figure. The first stage is relevant to ecologists but would it be better to identify the factors that people think are important in valuing species? Essentially arbitrary scores have been used in the method but the fundamental difficulty in a seemingly simple approach are the last two stages. There needs to be a rational basis for combining the scores and for multiplying the result otherwise a whole range of different methods could produce many different results. For those who are concerned about conservation, it could then be argued that calculation of monetary values or energy values is really just one step removed from ecological evaluation of the species or ecosystem. However, would the use of economic values, or alternatively monetary values, strengthen the case for conservation of a non-resource species? The answer may be no, but these methods of assessment are seemingly forced upon the conservationist in a world where monetary values and economic benefits are so powerful. What is worrying about the method for calculating notional monetary values is that it is misleadingly simple and it is not based on economic foundations.

The energy-based valuation method was briefly popular during the mid 1970s and although still supported by a few people it is equally problematic. Of the objections made by many economists, the most outstanding was that the assumptions required to make this a sensible theory just do not hold. Some economists would hold that it seems unrealistic to try and reduce all the attributes of an ecosystem to a single index (what does the figure derived mean and what is it a value of?). It is equally wrong to derive a value for energy by dividing GNP by primary energy input; the value of primary energy is measured by its economic price and the value of primary energy would typically by only about 5% of GNP (A. Ulph, personal communication).

Market values (although dependent on price, harvest rates, income levels etc.) are meaningful because they actually exist. But most species are not exchanged for money and that is where the problems begin and where the controversy lies. In the economist's attempts to give non-market values monetary figures, two popular and principal methods (among several methods) have evolved: replacement value and contingent valuation or willingness to pay. It seems that the replacement method has had some support among environmental economists but such a method is not really a way of valuing species and ecosystems. This is because it asks, given that we wish to conserve as best we can the species and ecosystem threatened by a development, is there a way by which an alternative (replacement) habitat can be constructed and what is the cost of doing that? The method does not tell us whether the value of the species justifies the cost of constructing a replacement habitat.

It would be naive to believe that arguments for conservation can be made effectively without considering alternative benefits. Economics concentrates on choice and presents a framework for thinking about and analysing very complex choices. Contingent Valuation (willingness to pay) seems to be logical but does it make sense for people with limited knowledge of biodiversity and ecology to contribute to Contingent Valuation Methods. Is there not a danger that those species which are perceived to serve certain needs will be favoured and conserved at the expense of other species? In research on information disclosure and endangered species valuation, it has been shown that information, appropriately selected, can influence the outcome of valuation studies, that is an individual's willingness to pay is influenced by the endangered status of the species and related biological information (Samples *et al.*, 1986). Surely, respondents must have relevant information about species before they can make statements about values, but what about all the species about which there is no or very little information? The role of information is very important here yet the information base for valuation of species is weak and there are many species for which we have little information about their role in life support systems (Randall, 1986).

Of course it does not follow that ecologists and conservationists must adopt existing methods. However, if they do not, then they must either develop alternatives (probably very difficult for most ecologists) or point out the limitations of existing methods. Clearly there is a continuing need for more collaborative research between economists, ecologists and conservationists. To this end it is good to see that during the last General Assembly of the IUCN in Perth (December 1990) it was decided to establish a working group on environmental assessment and resource economics (EARE) as part of the IUCN Commission on Environmental Strategy and Planning (CESP). One important objective is to provide a forum for dialogue between economists, ecologists and conservationists on concepts, methods and tools which can help to integrate ecological principles and conservation objectives into economic theory and planning.

REFERENCES

Aylward, B. (1991) *The Economic Value of Ecosystems: 3 – Biological Diversity.* London Environmental Economics Centre, Gatekeeper Series No. LEC GK 91–03.

Aylward, B. and Barbier, E.B. (1991) Valuing environmental functions in developing countries. Paper presented at the 'International Workshop on Ecology and Economics', Costa Rica, Centre Agronomico Tropical de Investigacion Y Enseñanza.

Brown, G. and Goldstein, J.H. (1984) A model for valuing endangered species. *Journal of Environmental Economics and Management*, 11, 303–9.

Costanza, R., Farber, S.C. and Maxwell, J. (1989) Valuation and management of wetland ecosystems. *Ecological Economics*, 1, 335–61.

de Groot, R.S. (in press) *Functions of Nature. Description and evaluation of the functions of nature as a tool in environmental planning, management and decision making.* Wolters-Nourdhoff, Groningen, The Netherlands.

Hair, J.D. (1988) The economics of conserving wetlands: a widening circle. Paper presented at workshop on economics, IUCN General Assembly, 4–5 February 1988, Costa Rica.

Hannon, B., Costanza, R. and Heredeen, R.A. (1986) Measures of energy cost and value in ecosystems. *Journal of Environmental Economics and Management*, 13, 391–401.

Helliwell, D.R. (1973) Priorities and values in nature conservation. *Journal of Environmental Management*, 1, 85–127.

Helliwell, D.R. (1985) *Planning for Nature Conservation*, Packard Publishing, Chichester.

Jacobs, M. (1991) *The Green Economy: Environment, sustainable development and the politics of the future.* Pluto Press, London.

Johansson, P. (1987) *The Economic Theory and Measurement of Environmental Benefits.* Cambridge University Press, Cambridge.

McNeely, J.A. (1988) *Economics and Biological Diversity.* IUCN, Gland.

Nijkamp, P. (1976) *Environmental Economics*. Martinus Nijhoff Social Science Division.

Nijkamp, P. (1977) *Theory and Application of Environmental Economics*. North-Holland, Amsterdam.

Nijkamp, P. and Verhage, C. (1976) Cost-benefit analysis and optimal control theory for environmental decisions: a case study of the Dollard estuary, in *Environment, Regional Science and Interregional Modelling* (eds M. Chatterji ands P. van Rompuy) Springer-Verlag, Berlin.

Nusslein, F. (1974) Der okonomische Wert der Wildproduktion in Europe (ohne Sowjetunion). *Z. Jagdwiss.*, **20**, 85–95.

Odum, H.T. and Odum, E.C. (1976) *Energy Basis for Man and Nature*. McGraw-Hill, New York.

Paul, C.K., Imhoff, M.L., Moore, D.G. and Sellman, A.M. (1989) Remote sensing of environmental change in the developing world, in *Changing the Global Environment: Perspectives on human involvement*, (eds D. B. Botkin, M.F. Caswell, J.E. Estes and A.A. Orio) Academic Press, London, pp. 203–13.

Pearce, D.W. and Markandya, A. (1989) *Environmental Policy Benefits: Monetary Valuation*. OECD, Paris.

Pearce, D., Markandya, A. and Barbier, E.B. (1989) *Blueprint for a Green Economy*, Earthscan, London.

Pearce, D., Barbier, E., Markandya, A., Barrett, S., Turner, R.K. and Swanson, T. (1991) *Blueprint 2: Greening the world economy*. Earthscan, London.

Randall, A. (1986) Human preferences, economics, and the preservation of species, in *The Preservation of Species* (ed. B.G. Norton) Princeton University Press, Princeton, NJ, pp. 79–109.

Samples, K.C., Dixon, J.A. and Gowen, M.M. (1986) Information disclosure and endangered species valuation. *Land Economics*, **62**, 306–12.

Schulz, W. (1985) Bessere Luft, Was ist sie uns wert? Eine Gesellschaftliche Bedarfs-Analyse auf der Basis Individueller Zahlungs-Bereitschaft, Technical University of Berlin, Germany.

WCED (1987) *Our Common Future*. Oxford University Press, Oxford.

Winpenny, J.T. (1991) *Values for the Environment: A guide to economic appraisal*. HMSO, London.

–4

Assessment for conservation and protection

4.1 INTRODUCTION

It is not realistic to spread all resources thinly in an attempt to protect and conserve all known endangered species. How do you choose between one species and another? Where there is much variation between populations of a species, what basis can be used to choose between populations? Is it best, for example to concentrate efforts and public funds on 'flagship' species (see p. 28) or is it better to focus attention on species which are of central ecological importance in a wildlife community? It could be argued, for example, that conservation efforts should be directed at keystone species (Paine, 1969) or key species (Lewis, 1978), that is those species on which the ecology of many other species depends. A third approach would be to identify the most taxonomically distinct species based on information of cladistic classifications; this approach is discussed below.

Alternatively conservation of some taxonomic groups (such as birds) may result in whole communities being conserved (p. 135). There are many individual species (including individual species of birds) which may justify special efforts to conserve them. In general, there has to be a selection of species for protection and conservation. Perhaps there are some general guidelines as to which categories of species might justify more conservation than others. For example, in connection with plants of New Zealand, Given (1981) suggested the following three categories:

1. Endemic taxa, in descending taxonomic significance, that is endemic families, endemic genera, endemic species;
2. Species in locations threatened by development;
3. Species from locations where the plant community is rich in endemics.

Locally rare species (see section 2.3 for account of rarity) could be overlooked if these three broad categories were always adopted and therefore it would seem better to treat each species or groups of species on their own merits, taking into account what is known about the ecology of the species and threats to the species.

4.2 SYSTEMATICS AND TAXONOMY AS A BASIS FOR ASSESSMENT

A common theme throughout this book is the dilemma posed by so many species to conserve yet there are so few resources for conservation efforts? A prerequisite for any assessment is identification and classification, for without knowing what the species is then it is impossible to make an assessment. That is, we need taxonomic skills to identify which populations are considered to be legitimate species.

Taxonomy and systematics provides us with more than just the name of a species, it provides us with information about the distinctiveness or uniqueness of a species (Gittleman and Pimm, 1991). Very sophisticated techniques are now available to help assess the taxonomic status of a species (e.g. see Wayne and Jenks, 1991). One popular way of measuring taxonomic distinctness is by cladistic relationships: the more taxonomically distinct the species, the more worthy of conservation.

Cladistic classifications are expressed as hierarchies and one method by which an index of taxonomic distinctiveness can be derived has been explained by Vane-Wright *et al.* (1991). In Figure 4.1, five related species (A,B,C,D and E) are considered. Each species belongs to a number of taxonomic groups (G), for example in Figure 4.1(a) species A belongs to the four groups AB, ABC, ABCD and ABCDE. The sum of the number of groups gives the score of 14. The number of groups to which each species belongs is the fundamental information for calculating the index (I) or basic taxonomic weights. For example the index for species A is 14/4 which is 3.5. Vane-Wright *et al.* suggest that it is convenient to standardize these values by dividing by the lowest score which gives the lowest ranking species a score of 1 (row W). Finally the standardized scores can be converted into percentages and that percentage value is an indication of the contribution of each species to the total diversity of the group of species. Thus in this example, species E is the most distinct and therefore the species worthy of most conservation effort.

Clearly, this method provides a basis for assessing conservation priorities from information on taxonomic distinctness but what is even more interesting and practical is that this method can be used for conservation evaluation of critical areas such as areas of endemism (p. 142). These areas would yield high scores that reflect the uniqueness of that species composition.

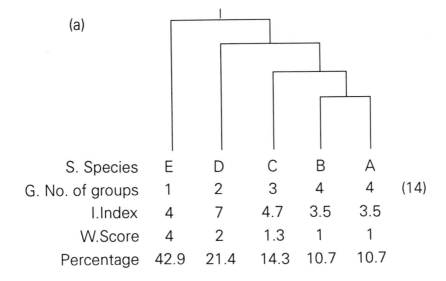

S. Species	E	D	C	B	A	
G. No. of groups	1	2	3	4	4	(14)
I.Index	4	7	4.7	3.5	3.5	
W.Score	4	2	1.3	1	1	
Percentage	42.9	21.4	14.3	10.7	10.7	

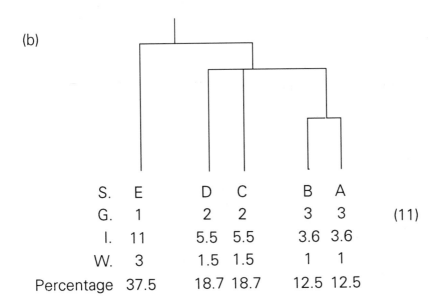

S.	E	D	C	B	A	
G.	1	2	2	3	3	(11)
I.	11	5.5	5.5	3.6	3.6	
W.	3	1.5	1.5	1	1	
Percentage	37.5	18.7	18.7	12.5	12.5	

Figure 4.1 Two examples showing the derivation of an index of taxonomic distinctness as proposed by Vane-Wright *et al.* (1991). These two examples are based on five hypothetical endemic species. See text for explanation.

4.3 WILDLIFE LAW

4.3.1 History and incentives

Wildlife law (Acts, Treaties or Conventions, Regulations and Directives) can play an important part in conservation. There is nothing new about wildlife law and there are examples of forestry conservation laws in Babylon which date back to 1900 BC. Lyster (1985) gives an example of land being set aside as a nature reserve as long ago as 1370 BC by the King of Egypt. Whereas some wildlife law deals with protected areas and special habitats (for example the Ramsar Convention, see p. 24) other wildlife law is directed at particular species or at groups of species. Whales, seals, polar bears, fish and birds are just some animals which have been the subject of international treaties or conventions. The incentives to establish wildlife law are varied. For example, support for protection of game species has and continues to have strong support from the game and hunting lobby. The conservation of non-game species such as song birds seems to have been supported with an economic argument. In the USA, the conservation of song birds as predators of insect pests has been advocated by a number of ornithologists (Smith, 1976). The fact that large sums of money are spent on observing or caring for wild birds and other forms of wildlife must be an important factor in the development of laws to protect wildlife. For example in 1974 about $170 million dollars were spent on wild bird food (Smith, 1976).

4.3.2 Wildlife conventions

International trade in wild plants and animals is a multi-billion dollar business much of which is illegal. The Convention on International Trade in Endangered species of Wild Flora and Fauna (CITES) has been particularly important and successful in its concerted efforts to prevent this exploitation of wild plants and animals. It is interesting in that it works as a protectionist treaty, by prohibiting, with exceptions, international trade in those species that are threatened (species listed in CITES Appendix I) and permitting controlled trading in some species (listed in CITES Appendix II). Appendix III of CITES provides a mechanism whereby a country with domestic legislation regulating the export of species not in the other appendices can seek support from other countries to enforce its own legislation.

CITES came into force in 1975 and although initially funded by the United Nations Environmental Programme, it is now funded by parties to the Convention (103 in 1990). CITES has a secretariat and is supported by staff and by consultants from around the world. The World Conservation Monitoring Unit at Cambridge has a particularly important role in maintaining databases and undertaking analyses for CITES. Despite the success of CITES, there has always been discussion about the criteria used for assessing a species for inclusion in one of the appendices. Initially the 450 species in Appendix I were the well-known, large endangered animals, such as the peregrine falcon (*Falco peregrinus*) and the cheetah (*Acinonyx jubatus*), but now the list has been greatly extended and includes many plant species. Appendix I of CITES includes those 'species threatened with extinction which are or may be affected by trade' but although there are criteria outlining the kind of information required to judge whether or not a species is threatened, CITES does not define what is meant by 'threatened with extinction'.

With 103 CITES parties, establishment of criteria for assessing whether or not a species is threatened would be impractical. There are therefore two levels at which CITES works. There are those lists agreed by the CITES Secretariat and there is a national level. At a national level, it would seem easier and indeed useful to have established criteria for assessing species in relation to wildlife legislation. The existence of national wildlife law directed at the conservation of particular species or taxonomic groups has come about largely as a result of public concern and the activities of conservation organizations. The selection of species has been based only partly on ecological information. However, implementation requires an input from both a management and scientific authority. Thus advice to the management authority is guided by ecological information. However, although 'ecological' input is involved in developing the Appendices, political and practical considerations cannot be avoided.

In Europe, one important wildlife Convention is the Berne Convention (The Convention on the Conservation of European Wildlife and Natural Habitats). This Convention, which came into force in 1982, was formulated over a period of six years and by 1986 a total of 17 European countries and the EEC were contracting parties to the convention. Developed through the Council of Europe, the aims are to conserve wild flora and fauna and their natural habitats, to promote cooperation between countries in their conservation efforts and to give particular attention to migratory, endangered and vulnerable species. This Convention has been important for conservation in Europe, for example it has been instrumental in the establishment of a European Network of 150 biogenetic reserves. There is an emphasis placed on the conservation of habitats and although no habitats are listed, the Berne Convention does contain lists (in the appendices) of plant and animal species for which various levels of protection have been agreed. Appendix 1, for example, is devoted to plants and appendices 2 and 3 to animals. Appendix 2 animals (e.g. the Dartford warbler, *Sylvia undata*, and the smooth snake, *Coronella austriaca*) are those thought to need special protection and appendix 3 animals (hedgehog, *Erinaceus europaeus* and the red squirrel *Sciurus vulgaris*) are those for which exploitation shall be regulated to keep the populations out of danger.

Curiously, there is no obvious explanation of the criteria used to select species in appendices 1, 2 and 3.

Lists of protected species need to be reviewed from time to time and the appendices in the Berne Convention are no exception. In 1990, for example, The Council of Europe in conjunction with the World Conservation Monitoring Centre (WCMC) was engaged in a revision of Appendix 1 of the Berne Convention. That revision was no casual affair and indeed it was the culmination of a two-year project conducted by the WCMC to develop a database on European plants which were considered to require international protection status. More than 100 botanists contributed to data on the status of about 500 plant species. As a result a new version of Appendix 1 was put before the Standing Committee of the Berne Convention early in 1991. Although the higher plants dominated the list, there were 26 bryophytes, 15 lichens, four fungi and even one species of algae. Amendments for other species can be made at any time by representations to the Berne Secretariat.

A proposed European Community Habitats and Species Directive will almost certainly be a milestone for conservation in Europe. The objective is to safeguard Europe's natural heritage by establishing a network of protected sites known as 'Natura 2000'. There are six annexes attached to the directive, the first two of which respectively list the habitat types and the species of flora and fauna eligible for protection. Annex III sets out the criteria for the selection of sites which may be identified as Sites of Community Importance and designated as Special Areas of Conservation (SACs). Annex IV lists species to be afforded strict protection and annex V contains species, of which the taking in the wild or exploitation should be the subject of management measures at the discretion of the member states. Annex VI broadly reflects the Berne Convention and is in keeping with existing UK legislation, in prohibiting certain means of capturing or killing mammals.

4.3.3 National wildlife law

Limited space prevents a comprehensive international approach and therefore only a few examples of national wildlife law have been selected from the USA and Britain. These two countries have been chosen because there is, overall, an interesting contrast: the US approach is largely a habitat-orientated approach related to the development of conservation and management programmes whereas in Britain the approach has been to look at countryside conservation and species conservation as two separate concepts. In the USA, laws to protect wild species and their habitats had their beginnings in 1908 when Congress established the National Bison Range where bison could be protected and managed. Many years later, The Wilderness Act of 1964 saw the introduction of a new approach in the conservation of natural or wilderness areas in North America and has led to much new legislation to protect natural features such as rivers and trails as well as plant and animal species. Interestingly, the concept of 'wilderness' as a land use category was first defined in a concise manner as a result of the 1964 Wilderness Act. An area of wilderness was defined in the Act as an area of undeveloped Federal land which retained its primaeval features, which did not have permanent improvements or human habitation and which is protected and managed so as to preserve its natural conditions and which

1. Generally appears to have been affected primarily by the forces of nature, with the imprint of work by humans substantially unnoticeable;
2. Has outstanding opportunities for solitude or a primitive and unconfined type of recreation;
3. Has at least 5000 acres of land or is of sufficient size to make practicable its preservation and use in an unimpaired condition;
4. May also contain ecological, geological, or other features of scientific, educational, scenic, or historical value.

The criteria used in the selection of wilderness areas and the rationale for management of wilderness areas are discussed in Chapter 6.

Laws aimed at the conservation of particular species commenced with the Endangered Species Preservation Act of 1966 which had the aim of providing a programme for the conservation and management of endangered wildlife. The Endangered Species Conservation Act of 1969 was enacted to prevent the importation of endangered species of fish and wildlife into the USA and also called for an international convention on trade in wildlife. These Acts did not prohibit the taking of endangered species and furthermore only those species threatened on a worldwide basis were protected.

A far more comprehensive act came later when the Endangered Species Act (ESA) was enacted in 1973. Despite its limitations (Rohlf, 1991), the Act has greatly helped the conservation of many species by way of preventing environmental degradation, habitat destruction and collecting. The main purpose of the Endangered Species Act of 1973 is to 'provide a means whereby the ecosystems upon which endangered species and threatened species depend may be conserved, to provide a programme for the conservation of such endangered species and threatened species . . .' The Endangered Species Act is particularly interesting from the point of view that it does not actually list species to be protected but provides a mechanism for the selection and thus the listing of species in need of protection. The Act also aims to conserve the appropriate ecosystem for the listed species. There have been a number of amendments to the 1973 Act and the Amendment of 1978 is particularly interesting in that one of the aims was to establish a committee to determine when exemptions could be granted.

Section 4 of the Endangered Species Act has a provision for the Secretary of the Interior to review the status of species that have been designated as requiring protection. Although anyone can submit evidence for the need to protect a species, the final decision is made either by the National Marine Fisheries Service or the Fish and Wildlife Service. Evidence that would be presented would include information on changes in population status, habitat condition and effects of trade.

More recently the USA has been discussing biodiversity legislation (HR 1268 Scheuer biodiversity bill). The aim is to establish fairly comprehensive laws for national conservation of biodiversity, the establishment of a national strategy to maintain biodiversity and the creation of a National Centre for Biological Diversity and Conservation Research.

In Britain, legislation relevant to the designation of protected areas or specially managed areas has included the Town and Country Planning Act 1947 (county council control over development), the National Parks and Access to the Countryside Act 1949 (National Parks, Areas of Outstanding Beauty, National Nature Reserves), the Countryside Act 1968 (National Parks and access to the open countryside), Nature Conservation Act 1972 (National Nature Reserves), Wildlife and Countryside Act 1981 (Sites of Special Scientific Interest), Agricultural Act 1986 (Environmentally Sensitive Areas).

Although Britain has enacted wildlife laws for the protection of various vertebrate groups (e.g. deer (1963), seals (1970), badgers (1973) and birds (1954)), it was not until 1975 when other 'animals' were included in the Conservation of Wild Creatures and Wild Plants Act (6 animals and 21 plants). In 1981 the Wildlife and Countryside Act was enacted. This Act contained extensive lists (schedules) of plants (Table 4.1) and animals. Although the Wildlife and Countryside Act includes schedules of protected species, the Act can also provide protection for habitats as the following quote (section 9(4) part a) from the Act implies: 'if any person intentionally (a) damages or destroys or obstructs access to, any structure or place which any wild animal included in schedule 5 uses for shelter or protection.'

The status of both protected and unprotected species may change and the 1981 Act has certain provisions for review of species at any time by the NCC (more recently the Joint Committee for Nature Conservation) or by the secretary of state. Also, the schedules in the Wildlife and Countryside Act are reviewed every five years with the addition and deletion of species. In general seven principles are used for the reviews:

Table 4.1 The Wildlife and Countryside Act 1981 (UK): Schedule 8. This is the original list of plants which were included in Schedule 8 (there have been revisions at each of the two subsequent quinquennial reviews). The British Red Data Book Threat Number and the IUCN Red Data Book Category has been included: E = endangered, V = vulnerable, R = rare.

Species†	Common Name	Threat Number	IUCN Category
Epipogium aphyllum	Ghost Orchid	13	E
Pyrus cordata	Plymouth Pear	13	E
Stachys germanica	Downy Woundwort	13	E
Bupleurum falcatum	Sickle-leaved Hare's-ear	12	E
Carex depauperata	Starved Wood-sedge	12	E
Cypripedium calceolus	Lady's Slipper	12	E
Damasonium alisma	Starfruit	12	E
Gnaphalium luteoalbum	Jersey Cudweed	12	E
Melampyrum arvense	Field Cow-wheat	12	E
Petroraghia nanteuilii	Childling Pink	12	E
Stachys alpina	Limestone Woundwort	12	E
Alisma gramineum	Ribbon leaved Water-plantain	11	E
Cephalanthera rubra	Red Helleborine	11	E
Cotoneaster integerrimus	Wild Cotoneaster	11	E
Eryngium campestre	Field Eryngo	11	V
Lactuca saligna	Least Lettuce	11	E
Orchis militaris	Military Orchid	11	V
Orobanche reticulata	Thistle Broomrape	11	E
Polygonatum verticillatum	Whorled Solomon's-seal	11	E
Veronica spicata	Spiked Speedwell	11	V
Althaea hirsuta	Rough Marsh-mallow	10	E
Artemisia campestris	Field Wormwood	10	E
Bupleurum baldense	Small Hare's-ear	10	E
Cyperus fuscus	Brown Galingale	10	E
Cystopteris dickieana	Dickie's Bladder Fern	10	E
Dianthus gratianopolitanus	Cheddar Pink	10	V
Diapensia lapponica	Diapensia	10	V
Himantoglossum hircinum	Lizard Orchid	10	V
Limonium recurvum	Sea Lavender	10	E
Orchis simia	Monkey Orchid	10	V
Scirpus triquetrus	Triangular Club-rush	10	E
Teucrium scordium	Water Germander	10	V
Viola persicifolia	Fen Violet	10	E
Allium sphaerocephalon	Round-headed Leek	9	V
Arenaria norvegica	Norwegian Sandwort	9	E*
Euphorbia peplis	Purple Spurge	9	E
Limonium paradoxum	Sea Lavender	9	E
Liparis loeselii	Fen Orchid	9	E
Minuartia stricta	Teesdale Sandwort	9	V

Table 4.1 continued

Ophrys fuciflora	Late Spider Orchid	9	V
Ophrys sphegodes	Early Spider Orchid	9	V
Polygonum maritimum	Sea Knotgrass	9	E
Potentilla rupestris	Rock Cinquefoil	9	V
Ranunculus ophioglossifolius	Adder's-tongue Spearwort	9	E
Trichomanes speciosum	Killarney Fern	9	E
Calamintha sylvatica	Wood Calamint	8	V
Lloydia serotina	Snowdon Lily	8	V
Phyllodoce caerulea	Blue Heath	8	V
Saxifraga cernua	Drooping Saxifrage	8	V
Scleranthus perennis	Perennial Knawel	8	E
Gentiana nivalis	Alpine Gentian	7	V
Gentiana verna	Spring Gentian	7	R
Gladiolus illyricus	Wild Gladiolus	7	V
Lychnis alpina	Alpine Catchfly	7	V
Woodsia ilvensis	Oblong Woodsia	7	V
Arenaria norvegica	Norwegian Sandwort	5	R*
Cicerbita alpina	Alpine Sow-thistle	5	R
Saxifraga cespitosa	Tufted Saxifrage	5	R
Woodsia alpina	Alpine Woodsia	5	R

*Two subspecies of Arenaria are listed in the British Red Data Book: Vascular Plants.
†This is the list as it was originally published but in a rearranged order.

1. Species already scheduled
2. Endangered or vulnerable species
3. Endemics
4. Species whose populations have undergone rapid decline in recent years
5. Species confined to threatened habitats
6. Species threatened in a European context
7. Species known only from a single site

Despite the provision for reviews and the use of these seven 'ecological' criteria, it would appear that the British Act does not have the flexibility of the Endangered Species Act.

4.4 PRIORITY AND RED LISTS

4.4.1 History

Three publications of the American Committee for International Wildlife Protection (one of which was *Extinct and Vanishing Mammals of the Western Hemisphere*) were important precursors of red data books (Fitter and Fitter, 1987). In the mid 1960s, the late Sir Peter Scott initiated the idea of IUCN Red Data Books so as to draw attention to the conservation needs of endangered species throughout the world. Originally the Red Data books were internal, private publications and expressed the work of one of the IUCN Commissions, The Species Survival Commission. The first Red Data Book to be made available for public subscription appeared in 1966. It was in loose-leaf form to allow regular updating and was the Red Data Book: Mammals (compiled by N. Simon) and was followed immediately by the Red Data Book: Birds (compiled by J. Vincent). In 1968 a Red Data Book for reptiles and amphibians appeared (compiled by R. Honegger). Plants were also to receive attention from IUCN and in 1970–71 the first version of the Red Data Book: Angiospermae (compiled by R. Melville) was published. This was so popular that it quickly became out of print and was then replaced by the IUCN Plant Red Data Book (Lucas and Synge, 1978).

Table 4.2 The categories of information for IUCN Red Data Book sheets

1 Common name	8 Habitat and ecology
2 Taxonomic name	9 Threats to survival
3 Order and Family	10 Conservation measures taken
4 IUCN threat category	11 Conservation measures proposed
5 Summarized overview	12 Captive breeding
6 Geographical distribution	13 Remarks
7 Population data and trends	14 References

The original idea of loose-leaf binding to allow updating of the synoptic information about each species was soon abandoned because of the costs of publication and the costs of the infrastructure required for distribution of new pages. The change in format was also indicative of the rate at which data were being collected and also of the realization that there were far more species in need of conservation than was ever thought possible. These books have not only served to focus attention on certain species but also provided a basis for gathering information on various species and promoting research in areas where information was required. It is because the main objectives of these books is to identify threats or causes of decline and not rarity or abundance *per se* that they have become essential for conservation planning and conservation in practice.

4.4.2 IUCN Red List Categories

The standard species data sheet for IUCN Red Data Books (Table 4.2) is a first step in the process of species assessment and assignment of the threat categories which have been defined as follows:

1. *Endangered.* Taxa (that is species or subspecies) in danger of extinction and whose future survival is unlikely if the factors continue to operate. Included are taxa whose numbers have been reduced to a critical level

or whose habitats have been drastically reduced so that they are thought to be in immediate danger of extinction. Examples of species in this category are given in Table 4.3.

2. *Vulnerable.* Taxa believed likely to move into the endangered category in the near future if the causal factors continue to operate. Included are taxa of which most or all the population are decreasing because of overexploitation, extensive destruction of the habitat or other environmental disturbance; taxa with populations that have been seriously depleted and whose ultimate security is not yet assured; and taxa with populations that are still abundant but are under threat from serious adverse factors throughout their range.

3. *Rare.* Taxa with small world populations that are not at present endangered or vulnerable, but are at risk. These taxa are usually localized within the restricted geographical areas of particular habitats or are thinly scattered over a more extensive range (see note about rarity on p. 38).

4. *Out of danger.* Taxa formerly included in one of the above categories but which are now considered relatively secure because effective conservation measures have been taken or the previous threat to their survival has been removed.

5. *Indeterminate.* Taxa that are suspected of belonging to one of the first three categories but for which insufficient information is currently available.

This simple priority classification was undertaken with three objectives in mind: first, it provided synoptic information on which to base conservation programmes; second, it has helped in the drafting of legislation; third, it is a good way of conveying information to the non-specialist. These objectives are most worthwhile but there has always been discussion about how the categories could be quantified and made more precise (Fitter and Fitter, 1987). One suggestion has come from Mace and Lande (1991) who proposed the following three categories of threat plus extinction:

1. Critical: 50% probability of extinction within 5 years or 2 generations whichever is the longer.
2. Endangered: 20% probability extinction within 20 years or 10 generations whichever is longer.
3. Vulnerable: 10% probability of extinction within 100 years.

The definitions are given here in general terms only and the authors provide more detailed criteria. Reminiscent of Seal's ISIS programme (p. 21), this proposal by Mace and Lande has the advantage of being based on population ecology but depends on good data from such studies.

Table 4.3 Examples of plants and animals in IUCN Red Data Book categories Endangered, Vulnerable and Rare

Endangered

Plants

Tetrataxis salicifolia. Only seven individuals of this tree are known, all from a single locality on Mauritius where it was found in 1975. Besides the grave danger of such a population being too small to maintain itself in the long term, the species is threatened by the invasion of exotics in its only locality. The site is unprotected and although a considerable distance from the nearest village, accidental damage to the remaining trees is therefore always a possibility.

Iris lortetii. Only four small populations are known at present, all in Israel. The original populations, never very large, were drastically reduced by commercial exploitation and are currently under threat from rapid habit destruction by afforestation and from the difficulties in restraining picking because of its exceptionally showy flowers.

Invertebrates

Anoglypta launcestonensis (Granulated Tasmanian Snail). This large land snail in a monotypic genus is endemic to Tasmania, Australia. It is found in an area of only 3000 ha and is very susceptible to land clearance. The creation of a reserve will be essential to guarantee its long-term survival.

Adelocosa anops (No-eyed Big-eyed Wolf Spider). This obligate subterranean blind spider is found in an extremely limited cave ecosystem on the Hawaiian island of Kauai, USA. The larger caves are under increasing pressure from visitors and surface development.

Mammals

Myrmecobius fasciatus (Numbat). Apparently confined to eucalyptus woodland in Western Australia though may conceivably still persist in arid inland regions further east. A survey in 1979 showed that a very marked decline in numbers had taken place in the 1970s and several former populations inhabiting reserves seemed to have disappeared.

Plecotus townsendii virginianus (Virginia Big-eared Bat). Cave-dwelling species found in eastern Kentucky, western Virginia and eastern West Virginia USA. Less than 5000 estimated to survive in 1979 and numbers are apparently declining with many caves being abandoned. Intolerant of human disturbance. Listed as endangered under the US Endangered Species Act of 1973.

Amphibians and Reptiles

Ambystoma matrodactylum croceum (Santa Cruz Long-toed Salamander). On the verge of total extinction. There appears to be little hope for this unique salamander, unless action is taken to preserve the breeding ponds and adjacent terrestrial habitat.

Eretmochelys imbricata (Hawksbill Turtle). Thinly scattered throughout the Atlantic, Pacific and Indian Oceans; also Gulf of Mexico and Caribbean. Formerly far more numerous throughout its range, which also included the Mediterranean.

Table 4.3 continued

Vulnerable
Plants

Phoenix theophrasti (Cretan Date Palm). A many-stemmed Mediterranean palm known only from five coastal localities. In much the largest locality at Vai it is threatened by tourists, by people camping under the trees and by cars driven into the centre of the grove, all of which prevents regeneration. It is also at risk from drainage schemes. In the other four localities it only occurs in very small numbers; in one of these it was evidently declining in 1967 and forming mere sparse low scrub; but in 1973 there was evidence of some regeneration.

Aeonium nobile. An attractive succulent of the Canary Islands which has declined due to clearing of its habitat for agriculture, to grazing particularly by goats, and to over-collection for horticultural use. Although frequent in places, it only occurs in a few localities.

Invertebrates

Formica spp. (Wood Ants of Europe). The wood ants of Europe include five closely related species. Highly valued for a number of services they perform in ecosystems, including control of pest insects and soil formation, all of these ants are declining over much of their range. Already subject to a great deal of conservation attention, they still require care to ensure that their numbers remain high enough to be of value.

Microchaetus spp. and *Tritogenia* spp. (South African Giant Earthworms). A group which includes the largest earthworm in the world, the South African endemic giant earthworms are characteristic of primary grasslands and savannahs. They are under threat from increased desertification caused by over-grazing and intensive cultivation.

Mammals

Arctocephalus townsendi (Guadalupe Fur Seal). Only known breeding colony is on Guadalupe Island off northwest Baja California. Commercial sealing drastically reduced the species and by the 1920s it was thought extinct. A colony was found in 1954. A 1977 population estimate gave 1300–1500 animals and censuses indicate a gradual increase. Only probable threat is increased human disturbance from a fishermen's camp but the impact of tourism also needs monitoring.

Canis lupus (Wolf). Formerly occurred in suitable habitats throughout much of the Northern Hemisphere, but now exterminated over large areas of its range and seriously depleted in much of the remainder. Numbers unknown. Habitat loss to human settlement and agriculture, stock predation, and a largely ungrounded fear of attacks upon humans have brought the Wolf into continual conflict with humans.

Amphibians and Reptiles

Python molurus (Indian or Burmese Python). Declining almost throughout the range of the species. In India, there has been heavy export of python skins for many years; in Hong Kong, the increasing human population has led to more disturbance of rural areas by urbanization and recreation, with consequent reduction of the scrubland habitat of the python; in addition, snakes are highly prized as food in Hong Kong and imported in numbers for that purpose from

Table 4.3 continued

China, but the size of the mainland population and its ability to withstand the pressure is uncertain.

Ambystoma tigrinum californiense (California Tiger Salamander). Found in Central Western California, from Sonoma and Sacramento Counties southward in the Central Valley to Kern County, and coastwise to Point Concepcion. It occurs in an area that is developing rapidly.

Rare
Plants
Papaver laestadianum. An endemic restricted to a small, mountainous area in Scandinavia. The species is plentiful in at least one of the four known localities. None of the populations are believed to be threatened at present and there are no reports of any decline. The localities are not easily accessible and are some distance from tourist routes.

Livistona mariae (Central Australian Cabbage Palm). Known only from an area of c. 60 km² in an Australian National Park. The palm is found in a very specialized, moist habitat of which there are only a few examples in the arid region where it occurs. Latz estimates that its population consists of around 1500 individuals of reproductive age and slightly more immature ones; it appears to have increased to some extent over the last 50 years.

Invertebrates
Birgus latro (Robber Crab). Probably the largest terrestrial arthropod in the world, this hermit crab is found throughout many of the islands of the Indo-Pacific. It has been intensively hunted by the local inhabitants in many places, and is reported to be extinct on a number of islands. Although detailed information is lacking there is a strong suggestion that it is declining in areas with heavy human populations. As well as being collected for food Robber Crabs are taken for sale as curios.

Dalla octomaculata (Eight-spotted Skipper). This is a rare member of the cloud forest Lepidoptera on the Caribbean slopes of the Central Cordillera of Costa Rica and western Panama. This community is particularly rich in endemic species. Some deforestation has occurred recently, although this is not yet as severe as that on the Pacific slopes of the range.

Mammals
Callimico goeldii (Goeldi's Marmoset). Disjunct and very sparse distribution in the Upper Amazonia rainforests, probably because of specialized habitat requirements. Occurs at low density in localized areas.

Phalanger interpositus (Stein's Cuscus). Scattered distribution in a narrow altitudinal band along the northern slopes of the central cordillera in New Guinea. Appears to be a relict form potentially very vulnerable to hunting and habitat destruction. Surveys are needed to determine more accurately distribution and to provide the basis for a conservation plan which should include establishment of reserves.

Amphibians and Reptiles
Leiopelma archeyi (Archey's Frog). Though of restricted range, locally abundant in suitable habitat of Coromandel Peninsula, New Zealand, at altitudes over 230 m.

Ctenotus lancelini (Lancelin Island Skink). Found only on a small island off the coast of Western Australia. Very little known about its ecology.

The basic methods of priority rating for the IUCN Red Data Books can equally well be used for small taxonomic groups. For example, Choudhury (1988) has used a basic but sensible priority rating to identify conservation of Indian Primates. Many of India's primates are threatened and all are of great biological interest; one of them, the hoolock or white-browed gibbon (*Hylobates hoolock*) is one of the closest relatives of humans. In the face of limited resources for conservation, a priority rating system was devised to identify those which were most endangered (Table 4.4). The degree of threat was mainly based on the extent of fragmentation of the habitat and populations. Scoring for taxonomic uniqueness was based on a continuum ranging from those which were the only member of a monotypic genus to members of large species groups.

Table 4.4 The priority ratings for conservation of Indian Primate species. From Choudhury (1988) with kind permission of the author

Species	Degree of threat	Taxonomic uniqueness	Association with other threatened forms	Total rating
Loris tardigradus	1	3	1	5
Nycticebus coucang	1	3	2	6
Macaca arctoides	4	2	2	8
Macaca assamensis	3	2	1	6
Macaca fascicularis	6	2	1	9
Macaca mulatta	1	1	1	3
Macaca nemestrina	3	2	2	7
Macaca radiata	1	1	1	3
Macaca silenus	5	2	2	9
Presbytis entellus	1	1	1	3
Presbytis geei	6	2	1	9
Presbytis johnii	5	2	2	9
Presbytis phayrei	6	2	2	10
Presbytis pileatus	1	2	2	5
Hylobates hoolock	5	2	2	9

The maximum overall score was 10 and on the basis of this priority ranking, Choudhury concluded that nine species need some special conservation attention. The effectiveness of this priority ranking will perhaps be tested in the future when resources may or may not be made available for the conservation of these primates. If wildlife sanctuaries are established for their conservation, then we can be sure that many other species of wildlife will be conserved along with the primates.

4.4.3 National red lists and threat numbers

The IUCN Red Data Books have served as an incentive for the publication of other Red Data Books or just plain Data Books. For example, the FAO (1986) has published a Databook on endangered tree and shrub species containing a list of species drawn up by a panel of experts on forest gene resources. The aim was to draw the attention of decision-makers, scientists and conservation organizations to the conservation needs of the 81 species threatened with extinction.

Several countries have now published red data books or red lists and there are in existence more than 100 national red data books and many more red lists. Countries which have produced red data books include Britain (Perring and Farrel, 1977; Shirt, 1987; Bratton, 1991), Russia (Borodin, 1978), Thailand (Humphrey and Bain, 1990) and New Zealand (Williams and Given, 1981; Wilson and Given, 1989). South Africa has a series of Red Data Books including those which deal with butterflies, reptiles and amphibians, terrestrial mammals, and birds. This series has been published in the series of South African Scientific Programmes Reports by the South African Foundation for Research and Development (address in Appendix A). Countries which have produced red lists of either plants or animals include Britain (Batten *et al.*, 1990), Africa (Hall *et al.*, 1980), Australia (Leigh and Boden, 1979), Europe (Council of Europe, 1977), North America (Fairbrothers and Hough, 1975; Kershaw and Morton, 1976) and Russia (Malyshev and Soboleuska, 1980). A comprehensive list of red data book and red lists for plants can be found in Synge (1981).

A few of these red data books attempt to rationalize the method for determining the degree of threat faced by a species or conservation needs of a species. For example, both Britain (Perring and Farrel, 1977) and Ireland (Curtis and McGough, 1988) have produced Red Data Books for vascular plants which incorporate an interesting method for calculating threat numbers. The use of 10 km squares (based on the national grid) in the Atlas of the British Flora to plot the distribution of plants was a major contribution to the method used for establishing threat numbers. To qualify for an entry in the Red Date Book, the species must be known from 15 or fewer 10 km squares. A total of 321 species (about 18% of Britain's flora) were selected for evaluation of threat (Table 4.5). Columns 1 and 2 contain information about the distribution of the species but do not contribute to the threat number. Column 3 refers to Great Britain and the first entry gives information about past and present distribution while the second entry contributes to the threat number (0 = 33% decline, 1 = 33–66% decline, 2 = 66% or greater decline). Column 4 gives the number of localities of the species in 1 km squares (0 = 16 or more localities, 1 = 10–15 localities, 2 = 6–9 localities, 3 = 3–5 localities, 4 = 1–2 localities. Column 5 is a subjective assessment of attractiveness of the species (0 =

not attractive, 1 = moderately attractive, 2 = highly attractive). Column 6 refers to extent of existing conservation (0 = more than 66% of localities on nature reserves, 1 = 33–66% localities on nature reserves, 2 = less than 33% on nature reserves). Columns 7 and 8 are subjective assessments of remoteness of the locality and the accessibility of the species (0 = not easily reached, 1 = moderately easily reached, 2 = easily reached). Column 9 is the threat number and is obtained by simply summing the values in

Table 4.5 A page from the British Red Data Book: Vascular Plants (1977 edition). From Perring and Farrel (1977) with kind permission of the authors

Species		Rate of Decline			Loci					
	Ireland (H)	Channel Isles (S)	Great Britain (GB)	Total	Attractiveness	Conservation	Remoteness	Accessibility	Threat number	IUCN category
	1	2	3	4	5	6	7	8	9	10
Minuartia stricta	–	–	1/1	0	1	4	1	1	1	2	9	V
Muscari atlanticum	–	–	10/17	1	17	0	2	2	2	2	9	V
Narcissus obvallaris	–	–	7/9	0	7	2	2	1	2	2	9	R
Neotinea maculata	19/32	–	1/1	0	1	4	2	2	0	1	9	R
Oenothera stricta	–	4/4	10/29	2	12	1	2	1	1	2	9	R
Ophrys fuciflora	–	–	4/6	1	10	1	2	1	2	2	9	R
Paeonia mascula	–	–	1/2	1	1	4	2	0	0	2	9	V
Phyllodoce caerulea	–	–	3/3	0	4	3	2	1	1	2	9	V
Polygonatum verticillatum	–	–	4/10	1	5	3	2	1	1	1	9	V
Polygonum maritimum	1/1	1/3	2/11	2	2	4	0	2	0	1	9	E
Potentilla rupestris	–	–	3/3	0	3	3	2	2	1	1	9	V
Ranunculus ophioglossifolius	–	0/1	2/4	1	2	4	0	1	1	2	9	E
Rhinanthus serotinus	–	–	5/68	2	10	1	1	2	1	2	9	V
Selinum carvifolia	–	–	2/5	1	2	4	0	1	2	1	9	V
Senecio cambrensis	–	–	5/5	0	6	2	1	2	2	2	9	R
Taraxacum acutum	–	–	2/2	0	2	4	0	2	1	2	9	V
Taraxacum austrinum	1/1	1/2	1/2	1	2	4	0	1	1	2	9	V
Taraxacum glaucinum	–	–	2/4	1	2	4	0	1	1	2	9	V
Tetragonolobus maritimus	–	–	9/9	0	9	2	1	2	2	2	9	R
Trichomanes speciosum	22/47	–	8/15	1	8	2	2	2	1	1	9	V
Trifolium bocconei	–	1/1	2/3	1	5	3	1	1	1	2	9	R
Valerianella rimosa	9/41	–	9/96	2	10	1	0	2	2	2	9	V
Veronica verna	–	–	1/8	2	8	2	0	1	2	2	9	E
Woodsia ilvensis	–	–	4/12	2	4	3	2	1	0	1	9	V

Table 4.6 The first 61 species in The British Red Data Book: Vascular Plants (Perring and Farrel, 1977)

Species	Common Name	Threat Number	IUCN Category
Alyssum alyssoides	Small Alison	13	E
Epipogium aphyllum	Ghost Orchid	13	E
Galium spurium	False Cleavers	13	E
Orobanche loricata	Oxtongue Broomrape	13	E
Pyrus cordata	Plymouth Pear	13	E
Senecio paludosus	Fen Ragwort	13	V
Stachys germanica	Downy Woundwort	13	E
Anthoxanthum puelii	Annual Vernal-grass	12	E
Armeria maritima subsp. *elongata*	Thrift	12	V
Bupleurum falcatum	Sickle-leaved Hare's-ear	12	E
Carex depauperata	Starved Wood-sedge	12	E
Caucalis platycarpos	Small Bur-parsley	12	E
Cyclamen hederifolium	Cyclamen	12	V
Cypripedium calceolus	Lady's-slipper	12	E
Damasonium alisma	Starfruit	12	E
Fumaria martinii		12	E
Galeopsis segetum	Downy Hemp-nettle	12	E
Geranium purpureum subsp. *forsteri*	Little-Robin	12	E
Gnaphalium luteoalbum	Jersey Cudweed	12	E
Iris spuria	Blue iris	12	V
Leucojum vernum	Spring Snowflake	12	V
Melampyrum arvense	Field Cow-wheat	12	E
Orobanche caryophyllacea	Bedstraw Broomrape	12	E
Petrorhagia nanteuilii	Childing Pink	12	E
Stachys alpina	Limestone Woundwort	12	E
Trifolium stellatum	Starry Clover	12	E
Agrostemma githago	Corncockle	11	E
Alisma gramineum	Ribbon-leaved Water-plantain	11	E
Campanula rapunculus	Rampion Bellflower	11	E
Cephalanthera rubra	Red Helleborine	11	E
Cotoneaster integerrimus	Wild Cotoneaster	11	E
Eryngium campestre	Field Eryngo	11	V
Geranium purpureum subsp. *purpureum*	Little-Robin	11	V
Hypericum linarifolium	Toad-leved St John's-wort	11	V
Lactuca saligna	Least Lettuce	11	E
Lonicera xylosteum	Fly Honeysuckle	11	V
Matthiola sinuata	Sea Stock	11	V
Orchis militaris	Military Orchid	11	V
Orobanche riticulata	Thistle Broomrape	11	E

Table 4.6 continued

Polygonatum verticillatum	Whorled Solomon's-seal	11	E
Rhinanthus serotinus	Greater Yellow-rattle	11	E
Scleranthus perennis subsp. *prostratus*	Perennial Knawel	11	E
Silene italica	Italian Catchfly	11	V
Veronica spicata subsp. *spicata*	Spiked Speedwell	11	V
Veronica triphyllos	Fingered Speedwell	11	E
Althaea hirsuta	Rough Marsh-mallow	10	E
Artemisia campestris	Field Wormwood	10	E
Bupleurum baldense	Small Hare's-ear	10	E
Centaurium tenuifolium	Slender Centaury	10	V
Cynoglossum germanicum	Green Hound's-tongue	10	V
Cyperus fuscus	Brown Galingale	10	E
Cystopteris dickieana	Dickie's Bladder-fern	10	E
Dianthus gratianopolitanus	Cheddar Pink	10	V
Diapensia lapponica	Diapensia	10	V
Himantoglossum hircinum	Lizard Orchid	10	V
Limonium recurvum		10	E
Linaria supina	Prostrate Toadflax	10	V
Mentha pulegium	Pennyroyal	10	V
Orchis simia	Monkey Orchid	10	V
Orobanche maritima	Carrot Broomrape	10	V
Pulicaria vulgaris	Small Fleabane	10	V

columns 3 to 8, the maximum threat number is 15. Column 10 is the IUCN Red Data Book category. A list of the first 61 plant species in the British Red Data Book is shown in Table 4.6 and it is interesting to compare that list with Table 4.1 and note the differences. For example, there are 26 species with threat numbers of 12 or 13 in the Red Data Book but only 11 species in the Wildlife and Countryside Act have a threat number of 12 or 13.

An obvious topic for debate is the subjective assessment of attractiveness, and perhaps the assessment of remoteness and accessibility. However, these criteria have almost certainly been instrumental in either the decline (if attractive) or the safeguarding (if inaccessible) of many plant species. If that were not convincing enough to justify the inclusion of these criteria in the calculation, the very simple scale of 0, 1 or 2 would seem not an unfair way of incorporating a subjective assessment of attractiveness, that is, a scale with three steps only is not going to produce much discrepancy or disagreement. Perhaps worthy of more serious discussion are the data on distribution and changes in distribution of a species. The data for the distribution maps, based on 10 km squares comes largely from volunteer efforts and the accuracy of the information depends on accurate identification of the species (as well as extensive surveys in each of the 10 km squares). It is possible that the number of 10 km squares in which the species has been reported to occur may not be accurate: lack of comprehensive surveys or incorrect identification could result in an inaccurate number of records. However, even if the error was as great as 10%, that would make very little difference to the threat number. Perhaps this mix of subjective data and numbers of squares gives a good balance for calculation of the degree of threat or conservation needs of a species.

The British Red Data Book on vascular plants was a rational attempt to evaluate the degree of threat and the conservation needs of a species and could therefore be used as a basis not only for focusing conservation efforts but also as a basis for assessing which species to include in the Wildlife and Countryside Act. Red Data Books also serve to draw attention to the growing number of threatened species but as Diamond (1988) points out the number of Red Data Books available can not cope with the growing number of threatened species, especially the many threatened species in the tropics. Just because a species is not in a Red Data Book does not mean that it is not threatened and therefore the idea of 'green lists' has been suggested by Imboden (1987). The use of 'green lists', that is those species which are known to be secure could be used to shift the burden of proof to those who maintain that all is well.

4.5 PRIORITIES FOR ENDANGERED SPECIES CONSERVATION

There is no doubt that red data books and red lists do play a valuable role in conservation either by drawing attention to the growing list of threatened species or by stimulating data collection. Drawing up lists of threatened species requires not only a method of establishing priorities for conservation but also requires surveys and monitoring so that threatened species can be recognized as such. Do we, for example, have effective monitoring procedures so that we know when a species previously rare has become vulnerable or endangered. The World Conservation Monitoring Centre has a particularly valuable role to play in so much as it acts as a centre for information gathering but in turn the WCMC must rely on networks of data gathering throughout the world. A 1984 survey of rare plant monitoring in the USA undertaken by Palmer (1987) revealed a considerable range of monitoring methods and most worrying of all was a widespread lack of review of the methods employed.

An important landmark in the evaluation and assessment of plant species in the USA was the Endangered Species Act 1973 (p. 87). At that time there was little information about the conservation status of plants in the USA as a whole and the Smithsonian Institute was asked to prepare an assessment of native ferns, gymnosperms and angiosperms. Rarity of the plants was an important criterion when drawing up the original list in

1974 then later through various consultations a more definitive list was published in 1978 (Ayensu, 1981). As well as national lists of threatened plants, almost every state in the USA has its own list of endangered, threatened and rare plants. It is perhaps not surprising, therefore, that almost every state has a different method for assessing the status of rare plant species. In response to this wide variation of criteria and methods of assessment used by various state, regional groups and government agencies in the USA, Ayensu (1981) suggested that it should be reasonable to develop a uniform set of assessment criteria for rare plants in the USA, based on observations of population structure and stability, environmental factors affecting populations, reproductive strategies and the role of natural succession. The following is an abbreviated list of recommended national assessment criteria (in reality they are qualities not criteria) as described by Ayensu (1981) in no order of priority:

1. Similarity of appearance to closely related species that may occur in the same area;
2. Vigour of populations, whether stable, increasing or declining;
3. Extent to which variability in numbers of individuals may be due to particular environmental factors that affect long-term population trends;
4. Frequency and density of populations;
5. Species biology, including reproduction, breeding systems, pollinators, dispersion, establishment of new individuals, maintenance and maturation classes;
6. Species in the area which may be highly competitive to the species under study;
7. Attractiveness of the species to collectors or commercial operators;
8. Location of fragmented populations;
9. Tolerance of the species to disturbance; ability to recolonize disturbed land;
10. Predation by herbivores, susceptibility to fire and other perturbations;
11. Stability of the preferred habitat;
12. Impact of natural succession of plant communities;
13. Habitat or genetic characteristics requiring conservation of more habitat than is usual;
14. Threats to the habitat by changes in land use;
15. Accessibility to the location whether protected or not protected;
16. Accessibility of plants to humans: steep cliff face compared to flat prairie sites;
17. Recommendations for 'critical habitat' (this refers to the Endangered Species Act of 1973 (p. 87)) designation;
18. Ownership of the habitat;

19. Economic impacts of conserving the species in a proposed development area;
20. Importance for possible chemical screening for medicines.

Some of these criteria are similar to those used in the British Red Data Book: Vascular Plants: namely, the criteria 2,4,7,8,15 and 16. A method of scoring and ranking species is described in the Red Data Book and Ayensu suggested that a numerical weighting of the value of the recommended criteria should be the next step.

Other methods and criteria for evaluating vulnerable plant and animal species status have been suggested by several authors including Ehrenfeld (1970), Sparrowe and Wright (1975), Adamus and Clough (1978), Soule (1983), Thompson (1984), Burke and Humphry (1978) and Millsap *et al.* (1990). Assessment of some of these methods have revealed limitations particularly with regard to those animal species not well studied. For example, Beiswenger (1986) applied several evaluation methods to the Wyoming toad (*Bufo hemiophrus baxteri*) with some rather disappointing results.

The Wyoming toad (Figure 4.2) declined markedly in the mid–1970s and only 25 individuals were found in 1980 and none were found two years later. The future of this species is in doubt. Beiswenger argued that the loss of this species may have been avoided if its decline had been detected earlier. On the basis of the identification scheme for extinction-prone species devised by Ehrenfeld (1970), only restricted distribution and intolerance of humans were applicable to the toad. The categories of large size, top predator, narrow habitat tolerance (stenoecious), economically important, hunted for market, migrates across international barriers, breeds in one or two vast aggregations, long gestation and behavioural idiosyncracies did not apply. The method proposed by Adamus and Clough (1978) requires fairly detailed knowledge about the species and lack of information on the Wyoming toad made it an unsuitable candidate. A method devised by Sparrowe and Wright (1975) gave a high score for the toad but over one-third of the score was based on the fact that data were not available.

Beiswenger (1986) concluded his assessment with five recommendations (very broad recommendations) to be included in early warning systems for identifying potentially endangered species as follows:

1. Identify characteristics which make species vulnerable and use those to identify those species which should be monitored;
2. Identify changes in population dynamics;
3. Maintain support for continued biological research (on the species suspected of becoming endangered);
4. Support and encourage the use of protected areas, especially those of high scientific value;

Figure 4.2 The Wyoming toad (*Bufo hemiophrys baxteri*). Photograph by David Withers kindly supplied by Ronald E. Beiswenger.

5. Establish monitoring programmes that are systematic and comprehensive yet cost-effective.

These are very broad recommendations and some need refinement; for example 'maintaining support for continued biological research' is a sound idea but impractical in general terms. The emphasis on monitoring is particularly important and there is a need for research on methods for monitoring long-term changes in ecosystems as well as species' populations (Spellerberg, 1991). The reference to identification of areas of high scientific value begs another question and that is discussed in Chapter 6.

The most recent and comprehensive research on priorities for animal species conservation has been undertaken by Millsap *et al.* (1990) who based their research on nearly 700 taxa native to Florida. Their objective was to provide a logical ranking of vertebrate taxa with respect to biological vulnerability, state of knowledge and management needs (Figure 4.3). The biological scores were sums of individual scores for seven variables that reflected different aspects of distribution, abundance and life history. The higher the score, the greater the vulnerability.

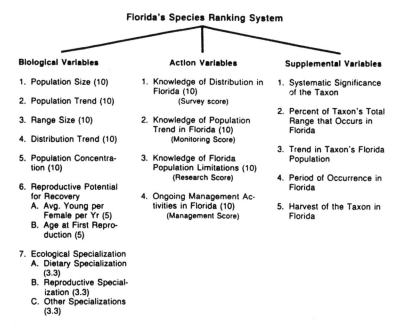

Florida's Species Ranking System

Biological Variables	Action Variables	Supplemental Variables
1. Population Size (10)	1. Knowledge of Distribution in Florida (10) (Survey score)	1. Systematic Significance of the Taxon
2. Population Trend (10)		2. Percent of Taxon's Total Range that Occurs in Florida
3. Range Size (10)	2. Knowledge of Population Trend in Florida (10) (Monitoring Score)	
4. Distribution Trend (10)		3. Trend in Taxon's Florida Population
5. Population Concentration (10)	3. Knowledge of Florida Population Limitations (10) (Research Score)	
		4. Period of Occurrence in Florida
6. Reproductive Potential for Recovery	4. Ongoing Management Activities in Florida (10) (Management Score)	5. Harvest of the Taxon in Florida
A. Avg. Young per Female per Yr (5)		
B. Age at First Reproduction (5)		
7. Ecological Specialization		
A. Dietary Specialization (3.3)		
B. Reproductive Specialization (3.3)		
C. Other Specializations (3.3)		

Figure 4.3 Florida's species ranking system as described by Millsap *et al.* (1990). Reproduced with permission of the authors and The Wildlife Society, USA.

It is not possible to devise a foolproof evaluation method for identification of threatened species that could be applied to all taxonomic groups in all kinds of ecosystems. The variation in life histories, ecologies and behaviour and the local circumstances have to be considered. Nevertheless Millsap *et al.* (1990) have shown that it is possible to devise an effective evaluation method for a range of vertebrate taxa (in Florida). Their method has since been used successfully by the Florida Game and Freshwater Fish Commission to allocate survey and monitoring efforts and to determine species conservation priorities. For example, some species such as the Eastern diamondback rattlesnake (*Crotalis horridus*), previously unprotected in Florida, was highlighted by their species ranking system as warranting better protection. It seems that Millsap *et al.* have devised a very successful evaluation method and one which could usefully be adopted by conservation agencies around the world.

4.6 CONCLUSIONS

In the few examples of wildlife law provided here, it is obvious that there is considerable variation in the strength of the law especially with regard to enforcement. For example, in general the conventions can threaten only disfavour between countries whereas directives must be enacted in existing legislation or in new legislation. General lack of enforcement seems to be a weakness of these conventions.

As one of the themes in this book is ecology, we should ask to what extent is ecology really used in the selection of species to be included in wildlife laws? It would seem that ecology has played a minor role in the selection of species for protection and this has been highlighted (Given, personal communication) in a survey of wild life law of 29 European countries. He found that lists of protected species contain not only endangered taxa but also attractive forms not necessarily in danger. He also found that some endangered species were omitted altogether.

In the absence of ecological data, an alternative approach is to protect all species but then to identify some exceptions. This approach has been used as a basis for protection of birds in the UK and is the basis of Britain's Endangered Species (Import & Export) Act (1976), which was particularly important in giving strength to the implementation of CITES agreements. Other comprehensive acts have been developed outside Britain and the USA; in New Zealand, for example, the protection of all species was used as a basis for the Wildlife Act of 1953. All wildlife (defined as mammals, birds, reptiles and amphibians) is protected with a few exceptions. Unprotected species and noxious animals are included in schedules appended to this law. Such an approach is certainly comprehensive and would seem to make the process of selection much easier. However, there still remains the problem of enforcement.

It is important to acknowledge the difference between criteria used to identify rarity or threat (conservation needs) and the need for protection. The argument can be put in two ways: (1) why should protected species be priorities for conservation when rarer species may be more threatened? (2) why should endangered species be protected if giving them statutory protection does not actually address the reason for the change in status? Thus a key criterion used in assessing the need for protection seems to be what direct conservation benefit will accrue from giving a species legislative protection.

The IUCN Red Data Books have played a very important role in directing attention to the needs of species throughout the world and the range of taxonomic groups continues to be extended despite problems such as those associated with the great diversity of invertebrates (Wells *et al.*, 1983). Deep sea organisms and pest species (including weeds) are two groups which could be the topics for future red data books. By way of contrast, some taxonomic groups have received what may be a disproportionate amount of attention. For example, detailed criteria have been used for identifying endangered plant species and for the drawing up of priority lists. The benefits of using many criteria and extensive information include the greater possibility of identifying reasons for a species decline and also provide an opportunity for assessing how best to help the species to recover.

The important applications of threat numbers seems to have been slow in gaining acceptance. It is a pity that the *British Red Data Book: Vascular Plants* has not been used more extensively when drawing up legislation for the protection of plants; that may change because British Red Data books are to be published on some non-flowering plant groups. Meanwhile the *Irish Red Data Book: Vascular Plants*, has been used as a basis for drawing up legislation for the protection of plants in Ireland. The use of these threat numbers may, in the future, extend to the Berne Convention provided that agreement can be reached about the use of an authoritative and standardized taxonomy. The publication *Flora Europaea* does provide a standard but it seems that not all countries are willing to accept that standard. Consequently it is difficult to apply the threat number methodology where there is disagreement about the taxonomic status of some species.

Although priority ranking and threat numbers do provide a good basis for conservation of species, there is much research to be undertaken on species evaluations. For example, how do you approach the conservation of a species made up of many populations with considerable variation between populations: which populations do you conserve. Conservation of genetic diversity must not be overshadowed by conservation of species. In a thought-provoking paper, Myers (1983) addresses some of these issues. In the knowledge that conservationists probably direct most of their efforts towards species known to science (see p. 3) and considering the work and early initiatives of the Species Survival Commission of the IUCN and the Endangered Species Office of the US Department of the Interior, Myers assesses several areas of research which might help to tackle the perplexing problem of how to prioritize species. Those areas of research and an indication of Myers' thoughts (only an indication because justice can not be done here to his paper) are as follows:

1. *Biological attributes.* For example, some species are predisposed to problems of survival because of isolated populations and specialized lifestyles. This has meant that island species have tended to receive special attention but twice as many continental forms as opposed to island forms are considered to be endangered.
2. *Ecological attributes.* For example, some species are essential for the survival of ecosystems (they have a keystone role) and their extinction leads to the extinction of many other species (domino effect). Also, indicator species (often rare) can play a valuable role in monitoring environmental quality (see examples in Spellerberg, 1991).
3. *Genetic attributes.* Endemic species command special attention because of their high levels of genetic variability but there is also the related question of critical minimum size for genetic reservoirs (see p. 4).
4. *Economic values.* Where are the promising candidates to be found and is it advisable for example to concentrate on edible diverse groups such

as the 13 000 species of legumes (see p. 16). What are the costs of the research needed to locate drugs for use in medicine?

5. *Cultural and aesthetic values.* Few endangered species generate concern, that is most endangered species are not furry or do not sport coloured feathers. Nevertheless there seem to be many indirect ways by which the public can be induced to support endangered wildlife of many forms.

6. *Species and exceptional value ecosystems.* Cost-effective conservation efforts can be achieved by directing efforts towards those areas which have high levels of species richness, thus much more could be achieved by conserving communities and ecosystems rather then species.

Myers' discourse clearly demonstrates that many of the methods used for priority ranking could be more effective if there was more research. More research requires time and therefore we need to adopt methods which hopefully will help to make conservation as cost-effective as possible. However, the research needed to devise better criteria and develop conceptual steps for evaluation of a species for protection may be more difficult than imagined. There are some very basic assumptions that have to be made including satisfactory taxonomic studies, accurate identification of the species and comprehensive surveys. Accurate identification and agreement on taxonomic distinctiveness, or which is a legitimate species, are perhaps greater problems with more widespread implications for conservation than has been supposed in the past (see for example Gittleman and Pimm, 1991; Wayne and Jenks, 1991). Certainly the need for more training in taxonomy and more taxonomists (as well as the adoption of standardized taxonomy) is a prerequisite for conservation.

Finally, there remains the need of enforcement of the law and security of protected areas. It is all very well having comprehensive wildlife law and red lists but both are of little use if the law is not enforced and red list species are not provided with secure protected areas.

Even more of a problem is the comparative importance of conservation or the priorities facing a country. There are many countries in the world where poverty and other desperate circumstances are going to put conservation low down in a list of priorities. This is not to suggest that conservation should be abandoned, on the contrary I am suggesting that realities should be recognized and where necessary there should be international cooperation to help fight poverty as well as threats faced by many species. This was recognized in the World Conservation Strategy and in the publication *Our Common Future* (WCED, 1987) which identified the need for economic development that is both ecologically and culturally sustainable. It is to be hoped that the successor to the World Conservation Strategy will pursue these problems.

REFERENCES

Adamus, P.R. and Clough, G.C. (1978) Evaluating species for protection in natural areas. *Biological Conservation*, **13**, 165–78.

Ayensu, E.S. (1981) Assessment of threatened plant species in the United States, in *The Biological Aspects of Rare Plant Conservation* (ed. H. Synge), John Wiley, Chichester, pp. 19–58.

Batten, L.A., Bibby, C.J., Clement, P., Elliott, G.D. and Porter, R.F. (1990) *Red Data Birds in Britain. Action for rare, threatened and important species.* T. & A.D. Poyser, London.

Beiswenger, R.E. (1986) An endangered species, the Wyoming toad *Bufo hemiophrys baxteri* – the importance of an early warning system. *Biological Conservation*, **37**, 59–71.

Borodin, A.M. (1978) *Red Data Book of the USSR.* Lesnaya Promyshlenost, Moscow.

Bratton, J.H. (1991) *British Red Data Books: 3, Invertebrates Other than Insects.* Joint Nature Conservation Committee, Peterborough.

Burke, R.L. and Humphrey, S.R. (1987) Rarity as a criterion for endangerment in Florida's fauna. *Oryx*, **21**, 97–102.

Choudhury, A. (1988) Priority ratings for conservation of Indian primates. *Oryx*, **22**, 89–94.

Council of Europe (1977) *List of Rare, Threatened and Endemic Plants in Europe.* Nature and Environment series No 14, Kew, Surrey.

Curtis, T.G.F. and McGough, H.N. (1988) *The Irish Red Data Book.* The Stationary Service, Dublin.

Diamond, J.M. (1988) Red books or green lists. *Nature*, **332**, 304–5.

Ehrenfeld, D.W. (1970) *Biological Conservation.* Holt Rinehart & Winston, New York.

Fairbrothers, D.E. and Hough, M.Y. (1975) *Rare or Endangered Vascular Plants of New Jersey.* New Jersey State Museum Scientific Notes 14.

FAO (1986) *Databook on Endangered Tree and Shrub Species and Provenances.* FAO, Rome.

Fitter, R. and Fitter, M. (eds) (1987) *The Road to Extinction: Problems of categorizing the status of taxa threatened with extinction.* IUCN in collaboration with UNEP, Gland.

Gittleman, J.L. and Pimm, S.L. (1991) Crying wolf in North America. *Nature*, **351**, 524–5.

Given, D.R. (1981) Threatened plants of New Zealand: documentation in a series of islands, in *The Biological Aspects of Rare Plant Conservation* (ed. H. Synge), John Wiley, Chichester, pp. 67–80.

Hall, A.V., de Winter, M., de Winter, B. and Oosterhout, S.A.M. (1980) *Threatened Plants of Southern Africa.* South Africa National Scientific Programmes Report No. 45, Cooperative Sciences Programme. Council for Scientific and Industrial Research, Pretoria.

Humphrey, S.R. and Bain, J.R. (1990) *Endangered Animals of Thailand.* Sandhill Crane Press, Gainesville.

Imboden, C. (1987) Green lists instead of Red Books, *World Birdwatch*, **9**(2), 2.

Kershaw, L.J. and Morton, J.K. (1976) Rare and potentially endangered species in the Canadian flora – a preliminary list of vascular plants. *Canadian Botanical Association Bulletin*, **9**, 26–30.

Leigh, J. and Boden, R. (1979) *Australian Flora in the Endangered Species Convention CITES*. Australian National Parks and Wildlife Service Special Publication 3.

Lewis, J.R. (1978) The implications of community structure for benthic monitoring studies. *Marine Pollution Bulletin*, **9**, 64–7.

Lucas, G. and Synge, H. (1978) *The IUCN Plant Red Data Book*, IUCN, Morges.

Lyster, S. (1985) *International Wildlife Law*. Grotius, Cambridge.

Mace, G.M. and Lande, R. (1991) Assessing extinction threats: towards a reevaluation of IUCN threatened species categories. *Conservation Biology*, **5**, 148–57.

Malyshev, V.L. and Sobeleuska, K.A. (1980) Rare and Endangered Plant Species of Siberia. Hayka. In Russian.

Millsap, B.A., Gore, A.G., Runde, D.E., and Cerulean, S.I. (1990) Setting priorities for the conservation of fish and wildlife species in Florida. *Wildlife Monographs*, **111**, 1–57.

Myers, N. (1983) A priority-ranking strategy for threatened species. *The Environmentalist*, **3**, 97–120.

Paine, R.T. (1969) The *Piaster – Tegula* interaction: prey catches, predator food preferences and intertidal community structure. *Ecology*, **50**, 950–61.

Palmer, M.E. (1987) A critical look at rare plant monitoring in the United States. *Biological Conservation*, **39**, 113–27.

Perring, F.H. and Farrel, L. (1977) *British Red Data Books: 1. Vascular Plants*. Royal Society for the Promotion of Nature Conservation, Nettleham.

Rohlf, D.J. (1991) Six biological reasons why the Endangered Species Act doesn't work – and what to do about it. *Conservation Biology*, **5**, 273–82.

Shirt, D.B. (1987) *British Red Data Books: 2. Insects*. Nature Conservancy Council, Peterborough.

Smith, R.L. (1976) Ecological genesis of endangered species: the philosophy of preservation. *Annual Review of Ecology and Systematics*, **7**, 33–55.

Soule, M.E. (1983) What do we really know about extinction? in *Genetics and Conservation* (eds. C.M. Schonewald-Cox *et al.*) Benjamin/Cummings, Menlo Park, CA, pp. 111–24.

Sparrowe, R.D. and Wright, H.M. (1975) Setting priorities for the endangered species programme. *Transactions of the 40th North American Wildlife and Natural Resource Conference*, pp. 142–55.

Spellerberg, I.F. (1991) *Monitoring Ecological Change*. Cambridge University Press, Cambridge.

Synge, H. (1981) *The Biological Aspects of Rare Plant Conservation*. John Wiley, Chichester.

Thompson, B.C. (1984) The numerical status rating system for nongame wildlife in Texas: attributes of public participation, in *Proceedings of the Workshop on Management of Nongame Species and Ecological Communities* (ed. W.C. McComb), Kentucky, USA, p. 59.

Vane-Wright, R.I., Humphries, C.J. and Williams, P.H. (1991) What to protect? – systematics and the agony of choice. *Biological Conservation*, **55**, 235–54.

van der Ploeg, S.W.F. and Vlijm, L. (1978) Ecological evaluation, nature conservation and land use planning with particular reference to methods used in The Netherlands. *Biological Conservation*, **14**, 197–221.

Wayne, R.K. and Jenks, S.M. (1991) Mitochondrial DNA analysis implying extensive hybridization of the endangered red wolf *Canis rufus*. *Nature*, **351**, 565–8.

Wells, S.M., Pyle, R.M. and Collins, N.M. (1983) *The IUCN Invertebrate Red Data Book*. IUCN, Gland, Switzerland.

Williams, G.R. and Given, D.R. (1981) *The Red Data Book of New Zealand*. Nature Conservation Council, Wellington.

Wilson, C.M. and Given, D.R. (1989) *Threatened Plants of New Zealand*. DSIR Publishing, Wellington.

WCED (1987) *Our Common Future*. Oxford University Press, Oxford.

-5

Ecological evaluation of biotic communities

5.1 INTRODUCTION

The intrinsic value (or the inherent value) of a biotic community can be described and calculated with the use of ecological and other criteria. There are however many different methods which have been developed for ecological evaluation and assessment of biotic communities. For example, some methods evaluate and then prioritize communities on the basis of their ecological, cultural and educational features. Other methods use characteristics of certain taxonomic groups (an indicator approach) to assess communities. Yet other methods attempt to adopt ecological approaches as much as possible and some combine several ecological criteria (including rarity, diversity, area, naturalness).

There are basically four scales of measurement which may be used in eological evaluations: nominal, ordinal, interval and ratio. The nominal scale is qualitative, that is it refers to mutually exclusive categories and has limited use in mathematical operations. The ordinal scale is used in the form of categories (e.g. low, medium and high) to which score classes are sometimes attributed. Allocation of score classes to ordinal values (e.g. 1 for low, 2 for medium) is arbitrary and subjective and the score classes should not therefore be summed. Values from interval (numerical distance between categories) and ratio scales (e.g. population density) are sometimes called scores (note that ordinal values are sometimes called score classes). Interval and ratio scales may involve assigning scores (e.g. 2 for 100+ species) which can then be used in mathematical operations (Smith and Theberge, 1987).

Some evaluations are based on several criteria and the result is then expressed in the form of an index. The use of indices to express the value of a biotic community has proved popular and not surprisingly there have been many suggestions as to the calculation of indices. Few evaluation methods have received international acceptance but there are exceptions. For example, an international and widely accepted method has been developed especially for the evaluation of avian habitats. Basically, if the area supports more than an agreed percentage of a biogeographical population of a particular species then that area is deemed worthy of special protection. The aim of this Chapter is to introduce some of the methods used for assessment and evaluation of biotic communities and to comment on their advantages and limitations.

5.2 PRIORITY RANKING AND HABITAT EVALUATION

5.2.1 Priority ranking

In 1974, Tans published a preliminary approach to the priority ranking of natural areas in Wisconsin, North America. His scheme was based on the simple allocation of points, the more points awarded the higher the rank. This evaluation method was designed to indicate the comparative ranking of an area when compared with other areas. The criteria were divided into four categories (biological characteristics, physical characteristics, degree of threat and availability for protective ownership). An outline of the various attributes within each category and the points awarded are shown in Table 5.1. In the biological category, up to 10 points could be awarded for the quality of the main biological features, up to 6 points for the measure of commonness, and up to 5 points for the measure of community diversity. The quality could be measured by the species richness, plant community structure and the integrity and extent of human interference. An area of high biological quality was considered to be a community type with no disturbance (an 'ideal' community). This method may be suitable for North America but not suitable for the landscapes of Britain and Europe where mankind's impact has been much greater.

Table 5.1 Priority ranking of natural areas. Criteria and point allocation suggested by Tans (1974). Reproduced with kind permission of the author

	Points
A. Biological Features	
1. *Quality*. This is based on richness of species, plant community structure, integrity, extent of disturbance, and naturalness.	
Highest Quality – area approaches the ideal community type: no disturbance or disturbance not visible.	10
High Quality – evidence of very minor disturbance	8
High Quality – at least one type of more obvious disturbance	6
Moderate Quality – one or more types of disturbance to community is obvious and community integrity threatened.	4
Low Quality – disturbance with resultant loss of the biotic community structure. May still have value as species habitats.	2

Table 5.1 continued

2. *Commonness.* A measure of importance of the area derived by evaluating the area, historical impacts, presence of rare or endangered species.

Very Uncommon – low area in presettlement vegetation and present vegetation, nearly complete conversion of type, restricted occurrence the presence of two or more rare or endangered species, or the only known location of a non-botanical feature.   6

Uncommon – moderate amount of type in presettlement vegetation and/or partial conversion of type.   4

Common – frequent to abundant in the present landscape, the type has increased since the advent of white settlement, or an adequate representation of the type within the scientific area system.   2

3. *Community Diversity.* The number of plant community types.

Great Diversity – four or more community types or features   5
Moderate Diversity – two or three types of features   3
No Diversity – single community type or feature   1

B. Physical Characteristics – Utility Value

1. *Area.* The minimum area for plant communities assumes adequate buffer zones but varies according to the community type.

2. *Buffer Zone.* This is adequate if it affords protection to an area.

Greatly exceeds minimum size, excellent buffer, no threat of encroachment.   8
Adequate size and buffer   6
Adequate size, inadequate buffer   4
Inadequate size, adequate buffer   2
Both inadequate   0

3. *Utility value.* Measured by the education value (at all levels of education).

Outstanding value – annually used by several schools or groups for both casual and structured activities; near metropolitan areas; extensive field station use or potential for extensive use.   4
Intermediate to high value   2, 3
Moderate value   1

C. Degree of Threat. Defined in terms of the area's security in respect to the maintenance of communities and natural features.

Threat is imminent; main features currently being developed or destroyed   10
Threat is imminent to portion of main features   8
Threat is moderate; development probable in future   6
Disturbance encroaching upon area   4
Little threat – destruction unlikely   2

Table 5.1 continued

D. Availability. An assessment of potential for protective ownership.

Available – offered as donation or owned by cooperating public agency	5
Available or at near appraisal cost, within an approved land acquisition boundary, or possible candidate for donation	4
Probably available at high cost	3
Availability in doubt – perhaps in time	2
Not available or available by condemnation	1

Tans (1974) suggested that for each area, the points allocated for quality, commonness, community diversity, size and buffer are summed and that the sum provides a basis for ranking. This method is based on score classes of ordinal values, that is ordinal values of low, medium and high are converted to scores of 1, 2 and 3 etc. This is a very convenient and most common method although strictly speaking it is invalid to sum or multiply score classes because each score class represents an arbitrarily chosen interval (see p. 113).

Following Tan's description of priority ranking, several other methods have been developed. Each person devising a new scheme is likely to be influenced by previous work and therefore although there are differences in criteria and methods of scoring, the results using different methods are sometimes very similar. The extent of agreement between three basic methods and a variation on one of those methods has, for example, been researched by Ogle (1981) in New Zealand (see O'Connor *et al.*, 1990 for a comprehensive review of evaluation for nature conservation in New Zealand). Ogle compared the methods used by Tans in America, methods of Wright in Britain (see p. 157) and two similar ranking schemes devised for habitats in New Zealand (G. Park, personal communication to Ogle). The results of the four ranks of 25 forested areas in Northland, New Zealand were analysed with Kendall's Coefficient of Concordance (*W*), corrected for tied rankings and it was found that differences in the order of ranking among the 25 sites achieved by the ranking schemes were not significant.

The similarity in the ranking of the three basic schemes might suggest that it does not matter which scheme is used. It would be unwise to support that conclusion because evaluation schemes should be modified for local ecological conditions and for data collection methods. It was partly for those reasons that Ogle proposed a revised scheme for ranking more than 360 forested areas in New Zealand which incorporated criteria previously used in other schemes (Ogle, 1982). Seven criteria were chosen, some entirely subjective and some easily quantified: representativeness, area, amount of forest near the habitat or degree of isolation, habitat diversity, habitat modification, number of indigenous forest bird species and rarity of other fauna (Table 5.2). Two indices for faunal rarity were included to give some recognition to smaller forest areas which are often rated poorly for birds but which contain large land snails (*Placostylus*, *Paryphanta*) gekos, frogs and bats.

Table 5.2 Criteria for ranking Northland's forests for wildlife in New Zealand. From Ogle (1981) with permission

A. Representativeness

	Potential	Moderate	High	Outstanding
A: *Representativeness* at local level	0	1	2	3
A: *Representativeness* at regional level	3	4	5	6
A: *Representativeness* at national level	6	7	8	9

B. Area (including scrub buffer zones)

>10 000 ha	= 5	50–249 ha	= 2
1000–9999 ha	= 4	10–49 ha	= 1
250–999 ha	= 3	<10 ha	= 0

C. Amount of forest near the habitat (i.e., degree of isolation)

An 'isolation index' is derived from the sum of scores for the proportion of forested land (measured in 1/64ths of the total area) within (a) a circle of 5 km radius; and concentric annuli within the ranges (b) 5–10 km, (c) 10–20 km, (d) 20–40 km, (e) 40–80 km of the habitat.

For ranking, an 'isolation index' (i.e. much forest nearby)	⩾ 120 = 2
'isolation index' 50–119	= 1
'isolation index' <50	= 0

D. Habitat diversity

Scores determined from evidence collected from maps, aerial photographs and Ministry of Works Land Resource Inventory Worksheets.

(a) within the habitat:
  (i) Relatively uniform = 0
  (ii) 2 or 3 forest/tall scrub types = 1
  (iii) >3 forest/tall scrub types *or* a small wetland, large stream etc. *within* the habitat = 2

(b) outside (but adjacent to) the habitat:
  (i) No natural area contiguous = 0
  (ii) Adjacent to unregistered or local-potential rated river, swamp, coast, etc. = 1
  (iii) Adjacent to higher rated river, swamp, coast, etc. = 2

Summary of Diversity Scores

Outside	Within 0	1	2
0	0	1	2
1	1	2	3
2	2	3	4

E. Habitat Modification

(a) >50% of forest area unmodified (all strata)   = 4

(b) <50% of forest area unmodified (all strata) *or* canopy intact with understorey lightly modified *or* intact mature 2° (secondary) forest   = 3

(c) >50% canopy intact understorey much modified *or* intact young 2° forest *or* intact tall shrubland on ridges, intact forest in gullies   = 2

(d) >50% canopy intact, understorey absent *or* <50% canopy intact, understorey much modified or absent *or* intact tall shrubland with emergent young trees   = 1

(e) much devastated canopy and understorey   = 0

F. Number of indigenous forest bird species

>9 species	= 2
5–9 species	= 1
<5 species	= 0

G. Rarity of indigenous birds

National rarity: or at least two regional rarities	= 4
Regional rarity:	= 3
Local rarity:	= 2
Good range of more common birds	= 1
Poorer range of more common birds	= 0

H. Rarity of other fauna

National rarity	= 2
Regional rarity	= 1
Nothing unusual	= 0

Maximum Total Score (Criteria A to H) = 32

The analysis of priority ranking undertaken by Ogle was initiated for a region by region survey of all natural areas with the aim of identifying those wildlife habitats of particular importance. Priority ranking is useful in those circumstances where countryside surveys are being undertaken but there are other occasions when the aim is to compare ecologically similar sites or a single evaluation of a particular area is desirable. Some of these evaluation methods are centred largely around ecological variables (Dony and Denholm, 1985) and some are of particular interest because of the use of ecological indices (Goldsmith, 1975; Ward and Evans, 1976; van der Ploeg and Vlijm, 1978; Buckley and Forbes, 1979; Peat, 1984).

A quantitative method for assessing, then ranking the conservation value of ecologically similar sites has been developed by Dony and Denholm (1985) based on species richness and rarity of vascular plant species of 23 small woods in England. They used three criteria:

1. Species richness (total number of vascular plant species in each woodland);
2. The proportion of species that are locally rare or uncommon (quantified on the basis of occurrence in tetrads (2km × 2km));
3. An index ignoring common species and weighted towards those rarest in the region as a whole.

Species locally uncommon ('selected' species) were defined as those recorded in 127 or below of the 371 tetrads (units of area 2 × 2 km) in that particular County (Bedfordshire). The proportion of locally rare species was calculated from the ratio of the number of 'selected' species to the total number of species in a woodland.

The relative rarity of a 'selected' species was quantified by assigning each a score (using an 'octave' system and thus the reason for selecting 127 tetrads or less). The score reflects the distribution by tetrads as follows:

Number of tetrad scores	1	2–3	4–7	8–15	16–31	32–63	64–127
Score	7	6	5	4	3	2	1

A plant rarity factor (PRF), the third measure used to assess and rank the woodlands, was derived from the sum of the scores for the 'selected' species on a site.

The difficulty in assessing the conservation values of ecologically similar sites is confounded by the area of the site because larger sites tend to have more species (see Spellerberg, 1991 for a review of biogeography and nature conservation). To overcome this problem, Dony and Denholm (1985) plotted each measure against the site area using both untransformed and log-plots and then selected the correlation that gave the best fit. Finally, site assessment values (AV) for each measure were calculated as the ratio of the observed value to the value predicted by the regression line for a site of the same area. That is, in one woodland there may have been a total of 79 species but the predicted number from a species–area plot (or log species–log area plot) might be 59. The AV therefore would be 1.3. Analysis of the ranks (Spearman's Rank Correlation Coefficient) of the woodlands derived from the three measures of species richness, revealed some consistent trends and some discrepancies. Although Dony and Denholm's method is a good first step for ranking ecologically similar areas, other factors may need to be considered.

5.2.2 Habitat evaluation

Habitat evaluation based on ecological science has been well researched in the USA, especially in connection with environmental impact assessments (see section 7.3) where the aim has been to ensure that appropriate consideration is given to wildlife in the decision-making process. At the same time, there has been considerable pressure for the use of standardized procedures for habitat evaluations, both for cost-effective reasons and for ease of communication of data both between and within organizations and professions. This pressure for standardization of inventories and evaluations was one of the reasons why the Habitat Evaluation Procedure (HEP) was developed (initially by the US Fish and Wildlife Service) for use in the evaluation of water and related land resource development projects (US Fish and Wildlife Service, 1980). First developed in 1976, HEP has since been modified after detailed assessments and there are now many descriptions of models for HEP (for example Lancia *et al.*, 1982).

In outline, the aim of HEP is to evaluate an area on the basis of the suitability of key habitat factors for certain species. That is, with detailed ecological information about a species, the characteristics of the habitat can be evaluated (using numerical rating schemes) on the basis of key habitat factors. The basic steps for the HEP are as follows:

1. The area being evaluated is divided into stands with relatively homogeneous cover types (e.g. mixed deciduous, secondary woodland, coppiced woodland).
2. Species are selected on the basis of guild representation for the evaluation (usually the species most sensitive to disturbance).
3. The extent of the habitat area for each species is estimated.
4. A Habitat Suitability Index (HSI) is calculated for each species for each stand in the evaluation area using ecological parameters (e.g. extent of canopy cover, successional stage). The HSI is defined as a value between 0 and 1 with the latter being the best quality of habitat in a defined area. The final aggregate value is an indication of the 'carrying capacity' of the area for that particular species.

An example of HEP based on the ecology of the marten (*Martes americana*) was used by Schamberger and Krohn (1982). On the basis of ecological information from literature on the marten, suitability index graphs were calculated: based on extent of tree canopy, extent of canopy composed of fir or spruce, and successional stage of stand (Figure 5.1).

The final HSI score is derived by aggregating the scores:

$$\text{HSI} = (V_1 \times V_2 \times V_3)^{1/3} \text{ (see Figure 5.1)}$$

The power of 1/3 was used because a simple geometric mean of the three values was desired (later developments of HEP have tended to use multivari-

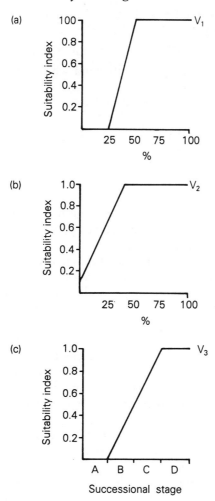

Figure 5.1 Habitat suitability index (HSI) graphs for winter cover required by the pine marten (*Martes americana*). (a) percentage tree canopy closure; (b) percentage of overstorey canopy closure composed of fir or spruce; (c) successional stage of stand where A = shrub seedling, B = pole sapling, C = young, D = mature or old growth. Redrawn from Schamberger and Krohn (1982).

ate regression). In the case of the marten, an area with 40% tree canopy closure ($V_1 = 0.8$), 24% canopy composed of fir or spruce ($V_2 = 0.4$) and a young successional stage ($V_3 = 0.7$) gives the following:

$$HSI = (0.8 \times 0.4 \times 0.7)^{0.33}$$
$$= 0.6$$

This example is made very simple by assuming only three ecological variables and one stand type. If there was more than one stand type then the final HSI would be calculated as follows:

$$\text{HSI} = (\Sigma(\text{HSI}_i \times a_i))/A$$

where HSI_i is the index for the ith stand which has area a_i and A is the sum of the stand areas (a_i).

Obviously the species selected and the extent of information used to select the ecological parameters is of key importance in HEP. The more sensitive species should be selected because they are going to be the first affected by perturbations such as physical disturbance and pollution. That selection could either be subjective or based on quantitative information about the species's tolerances to physical and chemical factors. Many models can be calculated for HEP but there has been a tendency to overlook exceptions. That is, although good ecological research can give very precise information about a species, sometimes plants and animals may survive without difficulty in what appears to be a less than satisfactory habitat. In other words, apparent survival at the time of surveying does not imply future success in the area and survival does not imply breeding success. At the same time we cannot ignore local ecotypes, especially those found among plants. We should always be alert to the exceptions and not be ruled by the expected.

Using several criteria for the evaluation of many sites (whether it be HEP or any other method) may lead to an accumulation of much data and this problem has sometimes led to the use of an index as a way of incorporating all the measures in just one value. Environmental indices have been popular in connection with measures of environmental quality in the USA (for example, Thomas, 1972) and ecological indices have been popular in Europe. For example Goldsmith's (1975) method of ecological evaluation was based on a case study which made use of a transect across three distinct land systems in Britain: unenclosed upland, enclosed cultivated land and enclosed flat land. Habitats within each land system were identified as arable and ley, permanent pasture, rough grazing, woodland, hedges and streams. The following variables were determined for each habitat type: 1, area (E); 2, rarity (R) for each habitat type calculated from $R = 100 - \%$ area per land system; 3, plant species richness (S) calculated from the number of flowering plant species in 400 m^2 plots; 4, animal species richness (V) based on the structure and stratification of the vegetation (number of vertical layers in the vegetation). An index of ecological value (IEV) for each kilometre square was calculated as follows:

$$IEV = \sum_{N}^{i=1} (E_i \times R_i \times V_i)$$

Another index similar to that used by Goldsmith is Pickering's index (see Marsh, 1978 and p. 214):

$$EV = \log_{10} \sum_{C=1}^{n} (V^3 \times e \times r \times s)$$

where EV is the index of ecological valuation, v is the number of vertical layers in the habitat, e is the extent of habitat in hectares, s is the total number of species and r is a rarity factor calculated as follows:

$r = 100 - C$
$C = e$ (extent of habitat in ha) / A (area of land in km^2)

The rationale behind the decision to cube the value obtained for v is explained on p. 216. It could be argued that there is no justification for this weighting and indeed if weighting is to be applied then rarity might be a better candidate. Also of interest is how to combine different criteria which have different weights? One possibility is by using the Saaty scale of relative importance and its application in multicriterion evaluations (p. 138).

A fundamental problem which must be addressed in all these ecological evaluations is that of definition of habitats (or biotopes) and scale of habitat classification (see also p. 52). Obviously habitats need to be defined in some way but even with a fairly definitive list of habitats, there may be difficulties in defining the limits of the habitat. Many of these evaluation methods do not take account of edges between habitats despite the fact that edges are sometimes very rich in both plants and animals. Ecological surveys have limitations of both time and money and therefore there has to be a balance between having large habitats with low survey costs but omitting small-scale diversity of habitats and a fine division which might include ditches and rotting wood as habitat categories (habitat diversity recorded but survey costs high).

One of the problems in ecological evaluations of communities is how best to measure and incorporate data on animals when there are so many different taxonomic groups which are not easy to sample (see Goldsmith's index above). By way of contract, and with appropriate taxonomic expertise, many species of flowering plants are easy to locate and very easy to record as long as the seasonal changes in visibility (annual plants, bulbs etc.) are borne in mind. Vegetation structure, plant species richness, diversity, composition and rarity are all relatively easy to quantify and all have been used in ecological evaluation methods (see, for example, van der Ploeg and Vlijm, 1978). Indeed, any of these variables can easily be incorporated in an ecological index or simple floristic index.

The ecological evaluation of the internationally and nationally rare limestone pavements in Britain (Figure 5.2) has lent itself to the use of a floristic index (FI). An index devised by Ward and Evans (1976) was based on species abundance ratings (1–3) with a weighting according to allocation to groups A to C (A, nationally rare species; B, nationally uncommon species or with a marked regional distribution; C, nationally common species). Data in the *Atlas of British Flora* (Perring and Walters, 1962) was used to quantify abundance. The FI was calculated as follows:

$$FI = 3 \Sigma a_A + 2 \Sigma a_B + \Sigma a_C$$

where the A, B and C species are thus weighted in the ratio 3:2:1, that is a nationally rare species is considered to be three times more valuable than a nationally common species.

A Compound Floristic Index (CFI) was also devised to provide a means of comparing groups of the limestone pavements as follows:

$$CFI = 3 \Sigma c_A + 2 \Sigma c_B + \Sigma c_C$$

where c is the average re-scaled abundance.

There are a number of ways of overcoming the problem of incorporating faunal characteristics in an evaluation, some based on indirect methods as proposed by Goldsmith, others based on easily observed taxonomic groups such as birds or butterflies (van der Ploeg and Vlijm, 1978), and others based on diversity of invertebrates collected by standardized, quantitative sampling (see also the criteria F, G and H in Table 5.2). For example Peat (1984) in an assessment of ecological evaluation of heathland communities proposed a Community Aggregate Score (CAS) as follows:

CAS = area (in km²) + floral species richness + diversity of invertebrates
+ rarity of the plant community

Floral species richness is simply the number of flowering plant species. Diversity is expressed as an index of diversity using invertebrates sampled with a D-Vac suction sampler over an area representing one square metre. There are many indices of species diversity (see p. 47 and review by Spellerberg, 1991) and the index employed by Peat was the Alpha Diversity Index (Williams, 1964). This index is expressed as follows:

$$S = \alpha \log_e (1 + N/\alpha)$$

where S is the total number of species, N is the total number of individuals and alpha is the diversity index. A nomogram (section 2.4.2) provides a quick and easy way of calculating the index from values of S and N.

Rarity of the plant community was calculated from the equation

$$R = 1 - P$$

where R is the rarity of the community and P is the proportion of that community within the area being evaluated.

Peat (1984) was interested in developing a simple index of ecological evaluation which could be obtained with a minimum of expertise and in as cost-effective manner as possible. He avoided weighting and converted

Figure 5.2 The internationally and nationally rare and threatened limestone pavements in England. These limestone pavements are rarer than tropical rain forests and are seriously threatened by the continued use of the stone for rock gardens. (a) Hill Castles (SD 994 681); (b) Southern Scales on Ingleborough (SD 744 767);

(c)

(c) *Asplenium trichomanes* at Feiger (SD 810 663). The Southern Scales on Inglebor-ough has been purchased by the RSNC and is managed as a nature reserve for them by the Yorkshire Wildlife Trust. Photographs taken in 1972/1973 and kindly provided by S.D. Ward (NCC).

each value to 1 so that the maximum value for the CAS was 4, thus making comparisons simple and easy to follow. As with all indices, they are convenient and can be used to summarize large sets of data. However, it must always be remembered that an index is simply a number and needs a reference for comparison. That is, the potential range of values that could be obtained by the use of a certain index should be quoted alongside the value being expressed.

A more serious problem is the reliability of the index, that is what range of values is obtained if several people independently calculate the index? Although calculation of area would present few problems, determining plant species richness is dependent on reliable identification and on the thoroughness of the survey for plants and time of year. The use of a standardized sampling technique as a basis for determining the species diversity of invertebrates minimizes variation between operators and makes little demand on time. Measurements of butterfly diversity or bird diversity may be affected by differences in recording techniques and therefore standardized methods have been developed for surveying and recording. Perhaps the most important variable with regard to interpretation of diversity of invertebrates is time of data collection (see Figure 2.1). Obviously different kinds of invertebrates would be collected at different times of the year and even at different times of the day (i.e. the catch depends on the activity of the species).

One of the reasons why Peat developed an index for ecological evaluation of heathlands was that lowland heathlands in Britain have become increasingly more damaged, greatly reduced in area, fragmented and isolated. This has resulted in many small remnants of heathland being threatened by further development. Fragmentation and isolation of biotic communities is a worldwide phenomenon and is one of the most important problems of current conservation (p. 12). It seems timely, therefore, that the evaluation of natural areas should consider extent of fragmentation and isolation.

Kent and Smart (1981) have developed a method for the assessment of semi-natural vegetation in Britain's agricultural landscapes which takes into consideration the fragmentation of the remnant habitats. The variables collected for each site were similar to those used by Ogle (see p. 116): site area, perimeter, distance to nearest other sites (measure of isolation), average size of the three nearest sites (an index of potential corridors or stepping stones), number of hedgerow contacts with a site, length of hedgerow within a site, number of intersections within the network of hedgerows, and the percentage of the perimeter of each site surrounded by grass leys or permanent pasture or arable crops. The variables describing the status of the site included habitat type (based on Elton and Miller's (1954) and Elton's (1966) categories – see Chapter 2 for explanation) and the plant species richness on the basis that the conservation value of a site generally increases with increasing structural diversity and increasing species richness.

Kent and Smart also proposed ordination and classification of the data so as to display the range of variation of habitat characteristics within each region and to determine groups of sites with similar characteristics. Although their detailed analysis is interesting from the point of view of identifying variables, there has to be a question about the amount of time that might be necessary for analysis. This would seem to apply when there appear to be alternative and less complex ways of identifying threatened remnants of vegetation without the necessity of going through the ordination and classification of data. What is important however, is that they have drawn attention to the need to overcome isolation of habitats by identifying groups of habitats, wildlife corridors and networks which could help to facilitate movement of individuals between habitats. In other words, they have drawn attention to the conservation needs of isolated habitats and the conservation value of wildlife corridors and networks (Spellerberg and Gaywood, 1993).

Rarity (or scarcity) is a common variable used in evaluation of natural areas and, although it can be quantified (see p. 38), some evaluations have introduced subjective assessments of the value of different levels of scarcity. For example, Sinden and Windsor (1981) in Australia developed a method for the selection of habitats based primarily on the idea of perceived satisfaction or utility provided by populations of species in different habitat. They used two measures in their evaluation: one is a measure of satisfaction or utility value of an area based on numbers of certain species; the second is a scarcity weighting based on comparisons of changes in the species

Table 5.3 Abundances of five species of birds in four hypothetical sites being considered for conservation. R = rare species, C = common species. From Sinden and Windsor (1981) with kind permission of the authors.

Species	*Sites*			
	A	B	C	D
Large sand dotterel (R) *Charadrius leschenaulti*	6	2	2	0
Knot (C) *Calidris* sp.	150	50	40	0
Caspian tern (R) *Hydroprogne caspia*	5	2	2	6
Common tern (C) *Sterna hirundo*	0	160	200	0
Sanderling (R) *Crocethia alba*	17	2	0	6

richness of certain species. The procedure for this method of estimating the value of wildlife preservation was illustrated with four hypothetical areas, each with five species of birds (Table 5.3) represented by different numbers of individuals or pairs.

Twenty four subjects were invited to take part in the estimation of values and all 24 were involved in conservation work. The utility values (or satisfaction) with respect to each species were obtained by asking each subject to rate the value of different levels of species abundance for each of the five bird species. In other words, levels of species abundance were converted to levels of utility by normalization in the form of a function graph. Function graphs are also employed in various methods of environmental assessment where the aim is to convert ecological variables into environmental quality (see review of environmental quality in Spellerberg,

1991). The weightings for different levels of abundance were obtained by asking subjects to allocate points between pairs of species given certain increases in abundance. The large sand dotterel was given a base weight of 1. For example, a subject would be asked to allocate 10 points between an increase of 0–10 pairs of large sand dotterels and the same increase in knot. This might result in 7 points to the large sand dotterel and 3 points for the knot indicating a preference for an increase in the former species. The scarcity weighting for the knot would therefore be 3/7 which is 0.43.

The selection of which sites to conserve can therefore be established by scoring the species using the weighting and utility value. An example of the scores for one subject and two sites is shown in Table 5.4. The scarcity weighting is multiplied by the utility value (based on the number of individuals) and a total for all species gives the value for the site.

Table 5.4 Calculation of the values of sites based on utility values and weighting for scarcity. From Sinden and Windsor (1981) with kind permission of the authors

		Site A		Site B	
Species	*Weight*	*Utility*	*Score*	*Utility*	*Score*
Large sand dotterel	1	3	3	1	1
Knot	0.4	11	4.4	8	3.2
Caspian tern	1	3	3	1	3
Common tern	0.2	0	0	10	2.0
Sanderling	0.6	3	1.8	1	0.6
Total for site			12.2		9.8

In the above examples, there is a trend towards the use of multicriteria techniques for ecological evaluations and the results are, in some cases, expressed in an index. A further example of a multicriterion technique (the Analytical Hierarchy Process) is described in section 5.4 together with a comparison with the use of the more commonly used indices. The benefits and disadvantages of such multicriterion techniques are discussed in section 5.8.

5.3 SYSTEMATICS AND TAXONOMIC GROUPS

5.3.1 Use of taxonomic scores

On p. 82 we saw how it is possible to measure taxonomic distinctiveness of a species and the contribution made by that species to the diversity of a taxonomic group. Here we look at how that idea can be taken further, in a very interesting manner, to identify important areas on the basis of taxonomic scores. In 1984, the concept of 'Critical Faunal Analysis' was introduced in a study of 158 species of milkweed butterflies (Danainae) by

Ackery and Vane-Wright. The authors attempted to identify the minimum number of areas which would support at least one population of every species. A total of 31 areas were identified in the following manner. All areas were placed in sequence on the basis of the number of endemic species present. The highest number of endemic species of the milkweed butterflies was found on the islands of Sulawesi (Indonesia). A programme of conservation on that island would therefore be aimed at 21% of the world species. A conservation programme on both Sulawesi and a small island of New Guinea could be directed at 29% of the world's milkweed butterfly fauna. It was concluded in that critical faunal analysis that 24 areas would be essential for the most narrow endemic species of milkweed butterflies but that 31 would be needed to protect all species.

The use of taxonomic scores for identifying areas of greatest importance for the conservation of centres of endemism is shown in Figure 5.3 (see also Figure 4.1). Imagine three areas, each supporting combinations of species of a certain taxonomic group. In one region there are species A, B and C (A is endemic), in another B, C and D, while in the third area there are species C,D and E (E is endemic). By using species richness, a variable commonly used in ecological evaluations, it would not be possible to distinguish between the three regions. Vane-Wright *et al.* (1991) have suggested that it would be possible to determine the most important areas based on taxonomic weightings in the following manner. On the basis of taxonomic distinctness (Figure 4.1), region A would score a total of 3.3 (35% of the total of 9), region B would score 46% and region C 78%. On this basis, region C would be the first choice and this would allow conservation of species C, D and E. Although region B scored 46%, a second priority would be region A because the combination of regions A and C give a score of 100%. This is because once we agree to region C, we need only be concerned about species A and B. On this basis region A can contribute 1+1 or 22% additional taxonomic diversity compared to only 11% which would be gained from using region B.

Measures of taxonomic diversity and the concept of critical faunas analysis are a useful basis for evaluation of areas for conservation because they identify not only centres of endemism but also the combinations of areas which would form the most useful basis for conservation. This approach shows very clearly that when assessing the status of taxonomic groups which include centres of endemism, it would be unrealistic to consider any one area in isolation. Systematics and taxonomy have much to offer species evaluations but the emphasis needs to be directed towards continuation of good taxonomic skills together with the exploitation of opportunities offered by recent advances in information technology.

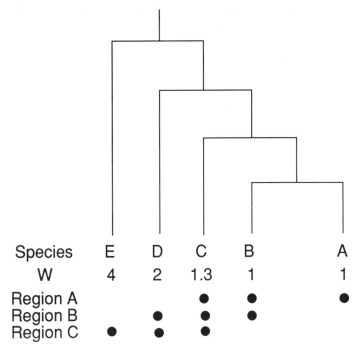

Species	E	D	C	B	A
W	4	2	1.3	1	1
Region A			●	●	●
Region B		●	●	●	
Region C	●	●	●		

Figure 5.3 The basis of priority area analysis for three regions with different combinations of related species and based on the taxonomic weighting. For example, region A supports three species, A, B, and C (A is endemic). W is the taxonomic weight. Region A scores 35%, region B scores 46% and region C 78%. See Figure 4.1 for calculation of the weighting. Modified from Vane-Wright *et al.* (1991).

5.3.2 Evaluation using taxonomic groups

Section 5.2 showed that an assessment of the conservation value of an area and the priority ranking of sites can be achieved by using ecological variables (either directly or indirectly) and subjective assessments of various features. Another approach is to use certain taxonomic groups as indicators of the value of a site. In many cases, invertebrate groups have been widely used for this purpose, partly because they appear to be easy to sample. Birds have been commonly used as indicators of the value of a site and therefore they are considered separately in section 5.4 together with assessments of habitats of birds.

An indicator species approach to evaluation of wildlife has been developed by Helliwell (1978, 1985). In one example he evaluated the wildlife on a number of farms and based the evaluation on the vascular plant species then the bird species. Conservation values of vascular plant species were calculated on the basis of measures of relative numbers in the study area, abundance in the region (taken as $10\,000$ km^2 in Helliwell, 1985), and occurrence in 10×10 km squares for the whole of the British Isles. The frequency of occurrence in 10×10 km squares was converted into a relative value from the equation:

$$\text{Relative value} = 1/\ (e^{-0.000676}\ y^2 + 0.1613y - 0.1606)\ 0.64_{xc}$$

where $e = 2.72$, y = occurrence (%), c (constant) = 0.00275.
Values can be read from a Table (see for example Helliwell, 1985 p. 104).

For example, if a species has a relative abundance of 100 000 in the study area, it occurs in 49% of the 10 km squares in the region and occurs in about 54.8% of 10 km squares in the British Isles. The relative value for 49% local occurrence is 3.8, the relative value for 54.8% national occurrence is 5.15 (obtained from Helliwell's Table of relative values). The conservation value for the species would then be:

Conservation value $= (100\ 000^{0.36} \times 3.8) + (100\ 000^{0.36} \times 5.15)$

$= (63.1 \times 3.8) + (63.8 \times 5.15)$

$=\ \ \ 564$

The power value of 0.36 was chosen by Helliwell in previous evaluations of vascular plants and is based on relative values from 1–360 ascribed to plants depending on their occurrence in 10 km squares in the British Isles. The conservation value of an area is a sum of the individual values for each plant species.

A similar scoring system was used by Helliwell for birds (Helliwell, 1978) in evaluations of wildlife on farms. However, he incorporated one further interesting but unsound element in the calculation. That is, the score for 'scarcity' was multiplied by the length of the bird in question. This was done 'to allow, to some extent, for the fact that an area with a population of two pairs of a large species could be regarded as being fully populated for that species, whereas a population of two pairs of a smaller species would not be at its theoretical maximum for that area'. This is not sound because it does not take into account species differences in space and resource requirements. For example, an area can be saturated with a few pairs of wrens (*Troglodytes troglodytes*) but have space for more magpies (*Pica pica*) even when there are several pairs.

There seem to be a number of assumptions in these calculations and one wonders if scores of 10 623 or 5876.2 for vascular plants and birds on farms can really help in assessments of the conservation value of the farms or any other areas. Clearly, relative abundance is an important factor in these calculations but there would seem to be simpler ways of incorporating abundance, rarity and scarcity into evaluation schemes.

Compared with some vertebrate groups, invertebrates have been used in only a limited manner in evaluations and classifications of natural areas.

Examples include work on Antarctic Islands (Usher and Edwards, 1986), natural areas in New Zealand (Ogle, 1981, see p. 116), fenlands in the east of England (Refseth, 1980; Luff *et al.*, 1989; Foster *et al.*, 1990), and aquatic habitats in England (Savage, 1982). The most common criteria used in conjunction with invertebrates have been various measures of species diversity (Chapter 2) and measures of similarity between sites. The underlying argument is that invertebrates such as beetles are easy to sample (pit-fall traps or D-vac suction traps) and that diversity and similarity indices are easy to calculate (but see Disney (1986) and also Chapter 2 for comments on ecological methods). This allows identification of the more diverse sites and the approach also provides a basis for the classification of the sites.

A classic example of the use of invertebrate taxonomic groups has been developed at Britain's Institute of Freshwater Ecology and the Freshwater Biological Association (FBA). A system has been developed using 'families' of invertebrates for monitoring freshwater communities. This classification system which is used to monitor levels of water quality is called 'RIVPACS' and it is based on the knowledge that the presence of certain taxonomic groups of invertebrates in rivers will require or are characteristic of certain physical and chemical conditions. The RIVPACS Programme commenced at the FBA in 1977 with the following objectives: to develop a classification system for unpolluted running water sites in Britain based on macroinvertebrate fauna, and to develop procedures for prediction of fauna which could be expected on the basis of environmental variables (Furse *et al.*, 1986; Wright *et al.*, 1989). The use of RIVPACS could be thought of as a way of measuring 'typicalness', one of the criteria used to assess sites of conservation interest (see pp. 185–189). That is, typicalness of a community is first determined on the basis of the presence of certain taxonomic groups, then the extent to which other communities bear any resemblance to the first community can be assessed.

A similar approach to that used for the RIVPACS scheme has been used for an extensive analysis and classification of water beetle assemblages in arable fenland on the east coast of England (Foster *et al.*, 1990; Foster, 1991). Their analyses arranged the data in a form which allows identification of important environmental variables and also objective identification of species assemblages (a similar approach has been used to quantify typicalness and representativeness: see p. 50). A point-scoring system (being developed by Foster) has been used to evaluate and rank wetland sites by the rarity status of the water beetle species present. The system proposed is based on a 'species quality score', an average of the points scored per species using a geometric progression doubling from 1 (commonest species) to 32 (rarest). The quality scores are standardized by dividing by the number of species present to give the 'Species Quality Score' as follows. Table 5.5 is an example based on the highest scoring site for one end-group arising out of the TWINSPAN classification. It is intended that this mean score per species should minimize any problems which may be caused by seasonality and by variation in collecting ability.

Table 5.5 Example of a Species Quality Score

Species score	Number	Points
1	12	12
2	5	10
4	5	20
8	0	0
16	1	16
32	0	0
Total	23	58

SQS (Species Quality Score) = 58/23 = 2.52

5.4 ECOLOGICAL EVALUATION OF AVIAN HABITATS

5.4.1 Popularity of birds in evaluation studies

Birds, together with vascular plants, are the most commonly used taxonomic groups in studies of assessment and evaluation of nature. A lot of effort seems to have been directed at the identification and protection of birds and their habitats. It has been argued that if bird communities are protected, then many other communities will be protected, and that evaluation of bird communities can be used as an indicator of the quality and conservation interest of the habitats (see p. 34). In general, birds are easily observed and many of them are familiar to people. The conspicuousness and popularity of birds would seem to be two reasons why there has been a considerable interest in studying and protecting birds.

For many species of birds, there has to be international cooperation for conservation of their habitat. Birds, like other animals, do not recognize political boundaries. Migratory birds require an international approach to their conservation. It is this international requirement for protection that has resulted in some very important and useful conventions and other forms of international law.

The evaluation of sites of ornithological interest was pursued with particular vigour during the 1970s at a time when the Ramsar Convention was established (p. 24). The Ramsar Convention, or at least the many discussions culminating in that Convention, focused much attention on the way sites of international importance, especially wetland sites, should be assessed. In addition to the specific criteria used for the selection of Ramsar sites, other criteria and methods have been used or proposed for assessing sites of importance to birds. For example, there has been much discussion about the use of indices (including multicriterion indices) and not surprisingly several methods for calculating an index of ornithological value have been suggested (for example, Nilsson and Nilsson, 1976; Fuller, 1980, Williams, 1980; Klopatek *et al.*, 1981; Harris *et al.*, 1983; Fuller and Langslow, 1986). Examples of indices are described here and the assessment and priority ranking of avian habitats is discussed.

5.4.2 Use of indices for assessing avian habitats

Over a period of ten years, from about 1976 to 1986 much attention has been directed at the potential use of indices for assessing avian habitats, particularly those habitats where there had been surveillance of bird populations for at least a year. One of the earliest proposals was for Swedish wetlands of ornithological interest (Nilsson and Nilsson, 1976). In their proposal, each species was given a conservation value inversely proportional to the species' estimated population in Western Europe and its breeding frequency in southern Sweden. The values for each pair of each species (which bred or probably bred in a locality in a particular year) were then summed to give the conservation value index. In their original index (Nilsson and Nilsson, 1976), log N_i was used but this was later rejected in favour of the following (Götmark *et al.*, 1986):

$$CV = \Sigma \, n_i \, /N_i$$

where CV is the index, n_i is the number of pairs of the ith species at a site and N_i is the population size of the ith species in Western Europe.

A very timely assessment of indices used to evaluate sites of ornithological interest was published by Götmark *et al.* (1986) who compared the use and application of five indices with their own method of evaluation (Box 5.1). Their own evaluation was not based on any particular formula or index but they did have quantitative information on rarity, species richness and population density of birds. Their own ranking of the sites was subjective, or more specifically, a result of mentally weighing different criteria and factors (Götmark *et al.*, 1986, Götmark, personal communication). The evaluations were carried out for two kinds of avian wetland habitats: grazed wet meadows (total of 15) and bogs (total of 45). The values for the ten best bogs are shown in Box 5.1.

Two of the five indices assessed by Götmark *et al.* were based, at least in part, on species diversity (Shannon–Weiner's Index and Simpson's Index). These criteria have been used in previous evaluations, for example by Bezzel and Reichholf (1974), Chanter and Owen (1976) and Harris *et al.* (1983). The evaluation method used by Harris *et al.* (1983) is described in detail on p. 156. Those evaluations which were based at least in part on diversity seemed to show a poor agreement with the evaluations undertaken by Götmark *et al.* (1986). Three other indices based largely on rarity

Box 5.1 The ranking of the ten most valuable bogs (based on bird census data) compared with the results of evaluations based on indices of diversity and rarity. The authors (Götmark *et al.*, 1986) evaluated 47 bogs, using their own method and five indices. The index values are given in brackets and because only 10 sites are shown here, the rank can be higher than 10. The equations for the indices are briefly explained below and further discussion about diversity indices is given on p. 212. b = index multiplied by 10. From Götmark *et al.* (1986) with kind permission.

Site	Authors' rank	Indices' rank				
		H'	θ	*CV*	V_k^b	B^b
A	1	8 (1·32)	1 (1·75)	1 (1429)	1 (1·51)	1 (7·98)
B	2	22 (0·98)	3 (1·29)	2 (668)	5 (0·59)	2 (3·83)
C	3	14·5 (1·12)	2 (1·48)	3 (425)	8 (0·44)	3 (2·80)
D	4	19·5 (1·01)	5 (1·18)	4 (381)	7 (0·47)	5 (1·94)
E	5	1 (1·70)	4 (1·28)	5 (303)	14 (0·24)	7 (1·54)
F	6	16 (1·06)	6 (1·07)	6 (197)	9 (0·33)	8 (1·03)
G	7	12·5 (1·18)	12·5 (0·91)	9 (117)	17 (0·22)	4 (2·06)
H	8	21 (0·99)	16 (0·85)	8 (130)	12 (0·26)	10 (0·71)
I	9	12·5 (1·18)	11 (0·92)	11·5 (106)	10 (0·30)	14 (0·59)
J	10	4 (1·47)	12·5 (0·91)	10 (107)	15 (0·23)	18 (0·53)

(1) Shannon–Weiner index of species diversity:

$$H' = - \sum_{i=1}^{s} p_i \, (\log P_i)$$

See p. 47 for further explanation.

(2) The θ index, which may be used 'when assessing areas that are candidates for conservation schemes' (Chanter and Owen, 1976):

$$\theta = \log (\beta N),$$

where N = total number of animals, and $\beta = 1 - \lambda$; λ is Simpson's index of diversity

$$\lambda = \frac{\sum_i n_i (n_i - 1)}{N(N - 1)},$$

where n_i = number of individuals of the ith species and N = total numbers (see alternative expression of this index on p. 47).

(3) *CV*, or 'Conservation value index', suggested by Nilsson and Nilsson (1976) as useful for the evaluation of wetlands as breeding habitats for birds:

$$CV = \sum_i \frac{n_i}{N_i}$$

where n_i = number of pairs of the ith species at a site, and
N_i = population size (pairs) of the ith species in western Europe.

Box 5.1 continued

(4) The V_k index, or 'the conservation value of the kth habitat' (Järvinen and Väisänen, 1978):

$$V_k = \Sigma_i \; V_{ik}$$

where V_{ik} is the conservation value of the ith species in habitat k

$$V_{ik} = \frac{n_i A}{N_i^2} \delta_{ik}$$

Here, n_i = population size of the ith species in a smaller region, such as a country or part of a country, A = area of this region, N_i = population size of the ith species in a larger region, such as several countries, and δ_{ik} = density of the ith species in habitat k.

(5) The B index ('the avian component in an overall ecological index'; Klopatek *et al.* 1981), for which the index value of the ith site is:

$$B_i = \mathop{\Sigma}_{j} \; F_{nj} F_{rj} F_{bj} F_{tj}$$

where the Fs are defined for the jth species as follows:

$$F_{nj} = 1 \cdot 0 - \frac{\text{no. of individuals of the } j\text{th species in the country}}{\text{total no. of individuals of all species in the country}}$$

$$F_{rj} = 1 \cdot 0 - \frac{\text{no. of individuals of the } j\text{th species in the region}}{\text{total no. of all species in the region}}$$

$$F_{bj} = \frac{\text{no. of the } j\text{th species in stratum } N}{\text{no. of the } j\text{th species in Bailey Ecosection } N}$$

$$F_{tj} = F_{nj} + F_{rj} + F_{bj}$$

(Nilsson and Nilsson, 1976; Jarvinen and Vaisanen, 1978; Klopatek *et al.*, 1981) showed a better agreement but were found to be influenced by size of the geographical area for which rarity was assessed.

The method proposed by Götmark *et al.* (1986) and their comments on previous methods of evaluation seems to have been well received by those engaged in ecological evaluation research and practice. One particular aspect highlighted by Gotmark *et al.* (1986), and quite rightly so, was the fact that many ecological indices are based on several non-independent criteria. That being the case, certain properties or qualities of a natural area could contribute to an evaluation through more than one criterion (note the comments about adding or multiplying score class for different criteria on p. 113). For example Ogle's evaluation of natural areas in New Zealand (p. 116) used representativeness (measured by field staff) which incorporates elements of area, habitat diversity etc. In addition, some criteria may be perceived as being relatively more important than others. Is it therefore better to use multicriterion techniques or is it more appropriate to use single criteria in isolation? Anselin *et al.* (1989) have pursued the idea of multicriterion techniques with an Analytic Hierarchy Process (AHP), originally developed by Saaty (1977), and compared their results with previous methods including that of Gotmark *et al.* (1986). The measurement scale according to Saaty (1977) is as follows:

1	=	Equal importance
3	=	Weak importance of one over the other
5	=	Essential or strong importance
7	=	Demonstrated importance
9	=	Absolute importance
2,4,6,8	=	intermediate values between two adjacent judgements

If activity *i* has one of the above assigned to it when compared with activity *j*, then *j* has the reciprocal value when compared with *i*.

In the hypothetical example given by Anselin *et al.* (1989) two ecologically similar sites are compared. The two sites combined support a total of 12 species of birds, seven species at each site (thus some bird species are common to both sites). There are three taxonomic groups of birds, each group represented by a number of species, passerines (P1-P5), waterbirds or Anseriformes (A1-A4) and waders or Charadriiformes (C1-C3). Site A has P1, P3, P4, A1, A2, C1, C2 and site 2 has P2, P3, P5, A3, A4, C2 and C3 (Box 5.2).

The overall important feature of the multicriterion technique AHP is that it converts the different criteria (such as area and naturalness) into comparable units and provides a basis for assessment of the relative weights or priorities of the different criteria derived from ecologically similar sites.

In the AHP, if F is the relative value of the site, the bird groups P,A and C are the criteria and the individual species of birds are the indicators. Alternatively, naturalness could be the criterion to which different indicators would contribute their relative weights or priority. In this example, it is assumed that of the three criteria, the value of P (the passerines) is of clear importance (therefore a value of 8 on the Saaty Scale) over both the other bird groups.

Next, Anselin *et al.* (1989) compared the three criteria, P, A and C in a 3 × 3 pairwise comparison matrix (Box 5.2). In that matrix, criterion P with criterion P is of equal importance and is therefore 1. Criterion P, has importance (score of 8) over the other criteria, therefore P with A is 8 and P with C is 8 (Box 5.2). With the Saaty scale, we have already noted that if *i* compared with *j* has a certain score, then *j* with *i* has a reciprocal score. Therefore A with P is 1/8 and C with P is 1/8.

The absolute weights from these relative pairwise weights can be obtained by finding the largest eigenvalue and associated eigenvector. In practice, however, the weights are unknown and must be derived (if done properly then it requires quite complex calculations). However, one rough method is to take the geometric mean of each row (the cube root of the product of the weights):

$$\sqrt[3]{(1 \times 8 \times 8)} \quad = \quad 4$$
$$\sqrt[3]{(1/8 \times 1 \times 1)} \quad = \quad 0.5$$
$$\sqrt[3]{(1/8 \times 1 \times 1)} \quad = \quad 0.5$$

The weights are scored as a fraction of 1 and so therefore in this example each result is then divided by 5 to give the weights or priority vectors of 0.8 (P), 0.1 (A) and 0.1 (C). In a similar manner, the indicators (that is the bird species) are compared for each criterion. The priority used by Anselin *et al.* (1989) was based on the national rarity of each species (expressed as number of breeding pairs). The rarer the species, the higher its priority. In the example given by Anselin *et al.* the species P1 is very rare, P4 and P5 are very common, P2 and P3 are less common. The species A1 is less common than A2, A3 and A4. For the species C, all are equal. The pairwise comparisons and derived weights for each species are shown in Box 5.2 where the Saaty values have been included. The weight of 0.590 for P1 is calculated from the geometric mean (5th root of the product of the 1,5,5,9,9).

The relative values of each site are then derived by the sum of multiplying the indicator (species) weights by the criteria weights. That is at site 1, the value is 0.48 (0.59 × 0.8) plus 0.13 (0.158 × 0.8) etc. (Box 5.2).

Finally, Anselin *et al.* applied this approach to the 47 bogs in Southern Sweden previously described by Götmark *et al.* (1986). The sites as ranked by Götmark *et al.* were compared by the ranking obtained by Anselin *et al.* Apart from the second and third sites, the ranking of the first six sites was found to be similar but there ended the agreement. The reason for this disagreement can be attributed mainly to the weights given to different groups of birds. If the same weightings are applied then there is a very close agreement between the ranks obtained from the two methods.

There is no doubt that in many evaluations and assessments, some criteria are considered to be more important than other criteria. That being the case, the multicriterion approach devised by Anselin *et al.* (1989) could have useful applications but it needs to be tested. There are, however, those who feel that a multicriterion approach is essentially flawed because some criteria are so similar and therefore contribute in a cumulative manner to the evaluation. That being true, only independent criteria should be included.

Box 5.2 A hypothetical application of a multicriterion technique proposed by Anselin *et al.* (1989).

In this example two sites (A,B) are compared and the criteria used were ornithological values of three groups of birds (passerines, waterbirds and waders). These groups are used as the three criteria in place of 'real' criteria such as area, naturalness and fragility. In reality, each of the two sites would consist of several habitats and the value of each habitat would be calculated using a selection of criteria such as area, naturalness and fragility. In this hypothetical example each criterion (group of birds) was quantified on the basis of their national rarity (expressed as number of breeding pairs). The rationale underlying the approach used by Anselin *et. al.* is to use Saaty scores of relative importance so that relative weights of different taxonomic groups (the criteria) can be used in a multicriterion approach.

Two sites (a,b): what are the relative values?

The bird taxa (criteria for evaluation) are: P(Passerines), A(waterbirds, Anseriformes), C(waders, Charadriiformes). There are 12 species in all, 5, Passerines, 4 waterbirds and 3 waders. Both sites (A,B) have 7 species each:

Site A.   Site B.
P1, P3, P4, A1, A2, C1, C2.   P2, P3, P5, A3, A4, C1, C3.

Of the three taxa (passerines, waterbirds and waders), it has been decided that passerines (P) have a value of clear importance (Saaty Value 8) over the other groups (A,C). This is like saying that of three criteria (area, naturalness and fragility) one, that is area, has a value of clear importance over the other two criteria. Application of the Saaty Scale is therefore as follows: C compared to P = 1/8; and A compared to P = 1/8

The relative weights of the three taxa are calculated as follows:

	P	A	C	weights	(calculated from the eigenvalue or as
P	1	8	8	0.8	an approximation from the cube root
A	1/8	1	1	0.1	of the product. The final weights are
C	1/8	1	1	0.1	expressed as a fraction of 1.0)

Taxon P has a weight of 0.8, A a weight of 0.1 and C a weight of 0.1.

Next, in a similar way as above, the weights of each species are calculated on the basis of national rarity. In this example, P1 is a very rare species, P4 and P5 are very common, P2 and P3 are less common. The species A1 is less common than A2, whereas A3 and A4 are equal. The waders have equal values. Pairwise comparisons for the Passerines would be undertaken as follows (note the application of the Saaty values 1, 5 and 9:

	P1	P2	P3	P4	P5	weights
P1	1	5	5	9	9	0.59
P2	1/5	1	1	4	4	0.158

Box 5.2 continued

P3	1/5	1	1	4	4	0.158
P4	1/9	1/4	1/4	1	1	0.047
P5	1/9	1/4	1/4	1	1	0.047

Pairwise comparisons would be undertaken for the waterbirds and the waders in the same manner.

Finally, the value of the two sites (A,B) can now be calculated by multiplying the relative weights of the individual species by the relative weights of the taxonomic group. Thus for site A which has species P1, P3 and P4, the first three values would be (0.8 × 0.59), (0.8 × 0.158) and (0.8 × 0.158). The value of each site is the sum of the taxonomic weights times the species weights:

Site A:
Value = 0.48 + 0.13 + 0.04 + 0.06 + 0.02 + 0.03 + 0.03 = 0.78

Site B:
Value =   = 0.36

5.4.3 International classification of wetland sites of ornithological interest

International assessment of wetlands and estuarine habitats for birds has been undertaken by the IUCN and by the International Waterfowl Research Bureau (IWRB). For example a MAR list of internationally important wetlands in Europe and North Africa has been drawn up, largely on the basis of ornithological data. The MAR Conference (sponsored by the IUCN, the International Council for Bird Preservation and The International Waterfowl Research Bureau) was held in 1962 (Saintes-Maries-de-la-Mer in the Camargue) to discuss internationally important wetlands. The MAR list published in 1965 (IUCN, 1965) includes 200 sites, all of which must reach certain conditions.

In 1974 at the International Conference on Wetlands and Waterfowl held at Heiligenhafen in Germany, three main sets of criteria were drawn up for the assessment of wetland sites (Box 5.3). Those criteria are now well accepted and are used by the participating countries. In Britain, for example, the protection of internationally important bird sites is well reviewed by Stroud *et al.* (1990).

The many discussions about MAR sites has led to common agreement of numerical criteria for assessment of wetlands of international importance. The main criteria are that the site should regularly support 1% or more of the regional population of a species or that it regularly supports more than 10 000 ducks or coot or at least 20 000 waders. Additional criteria are presence of endangered species, important resting sites during migration, and sites having educational value.

The proposal for a 1% criterion (an arbitrary figure without any biological basis) was formally adopted in 1974 at the International Conference on the Conservation of Wetlands and Waterfowl held at Heiligenhafen in Germany. A key statement was 'any site which regularly supports 1% or more of the birds of an entire species, or of a recognizable fly away or migratory population, or of any other clearly defined (biogeographical region) population should be regarded as being of international importance'. The 1% principle is now accepted widely throughout the world for assessing the importance of wetland sites for birds and is accepted not only by ornithologists and conservationists but also by policy makers and managers. Despite the fact that it has no real biological rationale, the simplicity of the principle seems to have made it acceptable.

So widely accepted is the 1% principle that is has also been applied to other groups of birds. For example, Lloyd (1984) devised a method for assessing the relative importance of seabird colonies around Ireland, partly on the basis of the 1% rule. An initial assessment of the Irish seabird colonies was made on the basis of the 1% principle together with information on rare species and additional conservation interests of the sites.

As well as being widely accepted and implemented, the 1% principle has been thoroughly discussed (Saeijs and Baptist, 1977; Lloyd, 1984; Fuller and Langslow, 1986). Although there seems to be little disagreement about the value of the principle, attention has been directed at the importance of management and security of sites after designation. Therefore, a fourth set of criteria applicable to conservation and management was agreed at the Heiligenhafen Conference in 1974 (Box 5.3). Another problem is the accuracy of census methods, particularly for those birds which are highly mobile and which occur in very large numbers. Annual assessments of bird populations are undertaken by many organizations throughout the world and many of the census methods have been scrutinized in detail (Prater and Lloyd, 1987).

5.4.4 Identifying and mapping centres of avian endemism

The International Council for Bird Preservation (ICBP) has a Biodiversity Project (initiated in 1988), the objective of which is to provide information for identification of key areas for conservation of global biodiversity using birds as indicators. An added bonus is that many key areas for avian endemism also support other groups of wildlife (Johnson, 1990). This project has established a unique database on the distribution of birds with estimated world ranges of less than 50 000 km². Each species is mapped using GIS and a multivariate statistical analysis was used to identify aggregations of species, termed Endemic Bird Areas. The preliminary analysis

Box 5.3 Criteria for assessing wetlands of international importance recommended by the International Conference on Wetlands and Waterfowl held at Heiligenhalfen, Germany, 1974.

---

1. Criteria pertaining to a wetland's importance to populations and species

A wetland should be considered internationally important if it:

(a) regularly supports 1% (being at least 100 individuals) of the flyway or biogeographical population of one species of waterfowl;

(b) regularly supports either 10 000 ducks, geese, and swans, or 10 000 coots or 20 000 waders;

(c) supports an appreciable number of an endangered species of plant or animal;

(d) is of special value for maintaining genetic and ecological diversity because of the quality and pecularities of its flora and fauna;

(e) plays a major role in its region as the habitat of plants and of aquatic and other animals of scientific or economic importance.

2. Criteria concerned with the selection of representative or unique wetlands

A wetland should be considered internationally important if it:

(a) is a representative example of a wetland community characteristic of its biogeographical region;

(b) exemplifies a critical stage or extreme in biological or hydromorphological processes;

(c) is an integral part of a peculiar physical feature.

3. Criteria concerned with the research, educational, or recreational values of wetlands

A wetland should be considered internationally important if it:

(a) is outstandingly important, well-situated, and well-equipped for scientific research and for education;

(b) is well studied and documented over many years and with a continuing programme of research of high value, regularly published and contributed to by the scientific community;

(c) offers special opportunities for promoting public understanding and appreciation of wetlands, open to people from several countries.

4. Criteria concerned with the practicality of conservation and management

Notwithstanding its fitness to be considered as internationally important on

Box 5.3 continued

one of the criteria set out under 1, 2 and 3 above a wetland should only be designated for inclusion in the list of the Ramsar Convention if it:

(a) is physically and administratively capable of being effectively conserved and managed;

(b) is free from the threat of a major impact of external pollution, hydrological interference, and land use or industrial practices;

(c) a wetland of national value only may, nevertheless, be considered of international importance if it forms a complex with another adjacent wetland of similar value across an international border.

has identified those regions of the world with the highest number of restricted-range bird species (Figure 5.4). One of the key questions now facing the ICBP is where conservation efforts should be directed. To answer this the ICBP has undertaken further analysis of its data. In the Philippines for example, 105 local endemic species of birds were divided into natural groups according to their ecologies and distributions. This enabled the ICBP to identify key habitat types and then it has been possible to decide which key habitat types can be conserved most effectively. This identification of key habitat types is similar to the concept of Critical Faunas Analysis developed by Ackery and Vane-Wright (1984). The ICBP is now reviewing the effectiveness of using birds as indicators for overall endemism.

5.5 ECOLOGICAL EVALUATION OF TEMPERATE WOODLANDS

5.5.1 Introduction

Temperate woodlands of North America, New Zealand, Britain, Europe and elsewhere have been much modified by man and his activities. Large areas have been cleared for agriculture and most remaining woods have sadly been damaged in some way. It is perhaps not surprising, therefore, that there has been much interest in how best to evaluate the surviving natural and semi-natural woodlands. Most woodland evaluations have sought to identify particular features of nature conservation value. Others have been directed towards amenity value and of course economic value (p. 71).

The most common features used to evaluate woodlands for nature conservation in Britain (for assessing nature conservation value) have been area of woodland, species richness of vascular plants, woodland structure and the presence and richness of habitats. Additional considerations have been the nature and extent of woodland management, features of the surrounding land and historical records. The first feature, area, has been favoured mainly because conservationists have, with regard to certain taxonomic groups, argued that larger woodlands support more species, an observation which has been mistakenly linked to theories of island biogeography. The ecology of woodlands should not be likened to the ecology of islands because whereas islands are surrounded by water, woodlands are an integral part of a patchwork-like landscape (Spellerberg, 1991). The use of species richness of vascular plants has been particularly popular and this approach has been extended to include weightings for rare species. Of course it should be remembered that, in general, species richness will increase with age and therefore a feature of ancient sites and woodlands is richness of ground flora. Structure refers to diversity and structural heterogeneity and in general the features contributing to high levels of structure include various management practices, coppicing, varied age classes and many layers. It is more precise, however, to consider structure of the stand (e.g. mixed, multilayered) or structure of the wood (different age-classes, open spaces and tree cover).

5.5.2 Use of indicator species

A notable contribution to evaluation of woodlands in the UK came in 1974 with the publication of Peterken's method for assessing woodland flora with indicator species (see p. 52 for account of biological indicators). Peterken explored the possibility of ranking a number of woods by counting the number of primary woodland vascular plant species present at each site. The rationale for using only woodland species (i.e. trees, shade-loving plants and woodland edge species) was that the evaluation should be relevant to the woodlands and should not be affected by ride species (rides are the equivalent of tracks or roads in woods or forests). This method of assessment can be qualified by counting only primary woodland ground flora species. A primary woodland is one which has been continuously wooded throughout the historical period and includes fragments of former natural forest cover and the species which are confined to such woodlands are the ancient woodland species (Table 5.6).

The advantages of the method are that it is simple and quantitative. It works because the number of species in each wood is determined by area, soil diversity and structural diversity, the main components of many woodland evaluations. The disadvantage is that one must first determine a list of primary woodland species, or at least a close proximation to it, which requires long experience. Most woods are important on grassland reserves (Peterken, 1991), and this has to be evaluated separately.

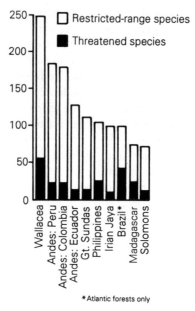

Figure 5.4 The regions of the world with highest restricted-range and threatened bird species. From Johnson (1990) with permission of the author.

5.5.3 Woodland evaluations based on several features

The use of indicator species as proposed by Peterken was later developed further and extended to include other features such as area, structure and woodland management (Massey *et al.*, 1977; Hinton, 1978; Smith, 1984). For example, Peterken and two colleagues (Massey *et al.*, 1977) combined a basic evaluation method with a woodland classification and extended the basic evaluation to include the total woodland flora except introduced species or invasive species. The woodland classification provided a basis for comparing like with like. It was also suggested that the evaluation could be extended to include subjective assessments of other features.

Classification of the woodlands into five types was based on the type of rocks underlying the woodland. This method avoided the problems of detailed surveys of species associations. Following on from classification, the various types of woodland are evaluated on the basis of woodland structure, species richness of woodland plants with a weighting for nationally rare species, area, and additional habitats. A simple, quantitative approach along these lines was applied to the woods of a whole county by Goodfellow and Peterken (1981).

Table 5.6 A list of primary woodland indicator vascular plant species. From Peterken (1974). This list has been modified in Peterken and Game (1984). Reproduced with kind permission of the author.

Group 1. Confined to primary woodland

Agropyron caninum	*Lathyrus montanus*
Aquilegia vulgaris	*Maianthemum bifolium*
Campanula trachelium	*Melampyrum pratense*
Carex laevigata	*Milium effusum*
C. pallescens	*Neottia nidus–avis*
C. pendula	*Oxalis acetosella*
C. strigosa	*Paris quadrifolia*
Chrysosplenium alternifolium	*Polystichum aculeatum*
Equisetum sylvaticum	*Scutellaria galericulata*
Lathraea squamaria	*Vicia sylvatica*

Group 2. Almost confined to primary woodland localities outside primary woods explicable survival on site, or (rarely) by planting

Carex remota	*Galium odoratum*
Chrysosplenium oppositifolium	*Luzula pilosa*
Galeobdolon luteum	*L. sylvatica*
	Melica uniflora

Group 3. Almost confined to primary woodland. Most colonize secondary woodlands very rarely

Anemone nemorosa	*Platanthera chlorantha*
Convallaria majalis	*Ranunculus auricomus*
Hypericum hirsutum	*Viola reichenbachiana*
Lysimachia nemorum	

Group 4. Most localities in primary woodland. Clear evidence that colonization can occur, but rarely

Adoxa moschatellina	*Dipsacus pilosus*
Allium ursinum	*Sanicula europaea*
Campanula latifolia	*Veronica montana*
Cardamine flexuosa	

Group 5. Most localities in primary woodland, but occur also in other long-established habitats. No evidence of a colonizing ability in either habitat

Calamagrostis canescens	*Orchis mascula*
Conopodium majus	*Primula vulgaris*
Geum rivale	

Group 6. Native trees and shrubs confined to primary woods and ancient, mixed hedges, except where (rarely) planted

Euonymus europaeus	*Sorbus torminalis*
Quercus petraea	*Tilia cordata*

5.5.4 Conservation priorities for woodlands

Comparing like with like seems to be very logical but there are many kinds of woodlands and in the face of limited resources there is a need to try and identify which types of woodlands (based on an agreed classification) should be the focus of most conservation efforts.

Remaining woodlands that are most difficult to recreate or those least modified by humans are the most important (old is best). Peterken (1977) suggested that five general types of woodlands were most important for nature conservation in Britain, namely:

1. Relicts of medieval wood-pasture systems;
2. Ancient high forest woods (mainly native pine woods);
3. Ancient coppice woods;
4. Ancient woods on inaccessible sites;
5. Woods formed by a long period of natural structural development.

Whereas this approach seems appropriate for woodlands which are small and isolated, it would not be appropriate in other temperate countries where woodland areas are much larger and less disturbed. Differences in management practices would also be an important consideration: in Britain for example coppicing is a valuable feature which continues to be practised but elsewhere, such as in Czechoslovakia, coppicing has been discontinued (Peterken, 1977, 1981).

5.5.5 Evaluation methods for temperate woodlands

Perhaps as is the case with the many other kinds of evaluation methods, putting methods into practice is more difficult than might be imagined. For example, defining the area being assessed is not straightforward for the many small, fragmented and modified woodlands in Britain. Also, the fact that some primary woods are contiguous with secondary woods does not make the delineation any easier.

Obviously the type of survey methods are an important consideration and there is an advantage in using standardized survey and recording methods (Chapter 2). There have been calls for standardization of the evaluation methods for woodlands, especially where like is being compared with like. Although there are good reasons for standardized methods when comparing similar woodland types, does it matter which method is used when evaluating different woodlands in one region? I think it does matter because some methods are not based on sound biological principles and tend to rely too heavily on an arbitrary selection of criteria.

5.6 ECOLOGICAL EVALUATION OF URBAN HABITATS

5.6.1 The status of urban nature conservation

Wildlife conservation in cities and towns tends to be associated with land-scaping and the restoration of derelict areas. Despite the growing interest in urban ecology, it is sad that nature conservation in urban areas has tended to attract less concern than conservation of natural or semi-natural areas because even unnatural plant assemblies on derelict sites and the animals they support (such as butterflies and birds) may be of both conservation and education interest. However, it is certainly true that most evaluation methods have been influenced by amenity and landscaping interests rather than nature conservation interests. Despite the emphasis on value to people rather than value to wildlife, urban ecology and urban nature conservation have flourished over the last ten years in many countries, including some parts of the USA, South Africa and Europe (Holland, Poland and Germany). The research on urban nature conservation during the 1980s seems to have generated an awareness that urban wildlife has many values including that of landscaping, enhancing amenity areas and reducing levels of airborne pollutants and noise (Sukopp *et al.*, 1990). That research has since been followed by very interesting work on the ecology of urban wildlife with particular reference to how fragmented and isolated wildlife populations survive in cities and towns. The role of urban linear features as both habitats and corridors for movement of wildlife has, in particular, received much attention (Spellerberg and Gaywood, 1993).

5.6.2 Urban evaluation methods

Of the few nature conservation evaluation methods which have been designed for urban environments, most have been developed as part of a planning process (Chapter 7). Other urban evaluation methods have been directed particularly at trees (p. 71) or woodlands (Berry, 1983; Domon *et al.*, 1986), which is perhaps not surprising because very often trees or woodlands are the sole remaining, obvious element of nature in a city.

In Germany where there has been much research on nature conservation in urban areas: the evaluation methods have been based on quite detailed botanical and zoological surveys and mapping (Brunner *et al.*, 1979; Sukopp *et al.*, 1980; Sukopp and Weiler, 1988). The use of both selective and comprehensive biotope mapping continues to be used in German and other European cities and one example of selective and rapid biotope mapping has been used by Wittig and Schreiber (1983). Working in Dusseldorf, they designed a 'quick' method for assessing the value of urban spaces for nature conservation. The biological parameters used for the evaluation (Box 5.4) are based not on botanical or zoological data but only on vegetation structure and therefore expertise in biology and ecology is not necessary. The evaluation is based on simultaneous consideration of the scores for each of the four parameters: D, period of development; A, area; R, rarity; H, function as a habitat (Box 5.4). A simultaneous consideration is adopted because the four scales are of different length and a simple sum of the scores would therefore not be acceptable. For particularly 'good' open spaces (in Dusseldorf) any of the following scores would be achieved (Wittig and Schreiber, 1983):

1. Score of 5 points in at least one of the scales D, R and H. Score 1 in A.
2. Score 4 in at least 2 of the three scales D, R and H. Score 1 in A
3. Score at least 3 points in each of D, R and H. Score 1 in A.
4. Score at least 5 in A and at least 2 in each of two other scales.

As, Wittig and Schreiber's method of evaluation was developed for a German town, it seems useful to see if the same method could be used in other localities. In 1983 the town of Worthing in Sussex (Southwest England) was chosen as a locality to test Wittig and Schreiber's method of evaluation of urban open spaces by Smith (1984). The methods of data collection needed only slight modification and on the whole were found to be suitable. The results showed that the H scale (function of the habitat) had a strong influence on the final result and led to some open spaces having high values largely on the basis of function of the habitat. By way of contrast, the R (rarity) scale appeared to have little or no effect on the final evaluation score. As a result it was decided to modify the H scale to make it more comparable with the other scales.

The R scale (rarity) was found not to be particularly suitable by Smith in her evaluation of a Sussex town and therefore another scale was devised called the extrinsic value scale (E) denoting value relative to other biotopes in the urban environment. This scale comprised two elements, typicalness (T) and value as a connecting biotope (C). Smith felt that typicalness (T) was a desirable element in this E scale because it would ensure that the best biotopes representing early successional stages were conserved. This helped to offset the bias of the D scale (of Wittig and Schreiber) towards older communities by helping to ensure that a representative sample of earlier seral stages were maintained and thus maintaining a wider range of habitats and species.

Function as a connecting biotope (C) was originally used as part of scale H by Wittig and Schreiber but it was concluded by Smith that this would be more useful as part of the E scale. That has important implications with respect to the idea of linear landscape features acting not only as habitats but also as corridors for dispersal of wildlife. Neither the scores of T (typicalness) or C (connecting biotope) were considered to be comparable with scores for H (function), D (period of development) and A (area), but Smith felt that cumulative effects should be incorporated in the evaluation.

Box 5.4 A rapid method developed by Wittig and Schreiber (1983) for the evaluation of urban spaces. Reproduced with permission of R. Wittig.

The period of development (D) means taking into account the length of time that would be necessary for the same community to have established itself in another location.

Scale of assessment D:

$$D0 = 1–2 \text{ years}$$
$$D1 = 2–5 \text{ years}$$
$$D2 = 5–10 \text{ years}$$
$$D3 = 10–20 \text{ years}$$
$$D4 = 20–50 \text{ years}$$
$$D5 = 50–100 \text{ years}$$

Area size (A)

The larger the area, the greater is the chance of a community of numerous species developing in it. For this reason the area size is used as a criterion for the value of the biotope.

Scale of assessment A:

$$A0 = 0.1 \text{ ha}$$
$$A1 = 0.1–<1\text{ha}$$
$$A2 = 1–<5 \text{ ha}$$
$$A3 = 5–<10 \text{ ha}$$
$$A4 = 10–<25\text{ha}$$
$$A5 = 25–<50 \text{ ha}$$
$$A6 = 50–<100\text{ha}$$
$$A7 = 100–<200 \text{ ha}$$
$$A8 = \geqslant 200 \text{ ha}$$

Rarity (R)

For the assessment of the rarity of the urban biotopes the following scale was applied:

R0 = many similar biotopes occurring in the built-up urban area, the nearest equivalent less than 500 m away

R1 = several similar biotopes in the built-up area, the nearest equivalent 500–1000 m away

R2 = several similar biotopes in the built-up urban aea, the nearest equivalent 1000–2000 m away

R3 = the nearest equivalent biotope over 2 km away, or only approx. 5–10 corresponding biotopes in the built-up urban area

R4 = only 1–4 corresponding biotopes in the built-up urban area

R5 = no equivalent biotopes in the built-up urban area, but existing in remaining urban area or within adjoining communes (up to approx. 5 km from the city boundaries)

Box 5.4 continued

> R6 = no equivalent biotopes in the urban area and local surroundings (see above) but several (more than 5) in the total administrative area (metropolitan county)
>
> R7 = no more than five equivalent biotopes in the whole metropolitan county area.
>
> Biotopes are considered of greater value if they fulfil the functions of the biotope under comparison better or possess additional functions. For example, a park can fulfil all the biotope functions of a churchyard, often to a greater degree, whereas a churchyard cannot fulfil all the functions of an optimally developed park.
>
> **The function as habitat (H)**
>
> Within the framework of the Dusseldorf study, only the vegetation structure was considered in detail. The following scale of assessment was developed:
>
> H0 = almost exclusively grass or trodden-area communities
> H1 = Almost exclusively a uniform vegetation structure, but not grass or trodden-area community
> H2 = two different vegetation structures
> (H3, H4, etc. correspondingly).
>
> The following were considered as different vegetation structures:
> Wood/wood-like parks
> Row(s) of trees
> Group(s) of trees
> Single trees
> Hedges and shrubbery
> Areas of tall herbaceous plants
> Hay meadow
> Pasture
> Park grass/ornamental lawn
> Trodden-area communities
> Therophyte areas
> (Semi) ruderal xerothermic grassland
> Wall communities
> Reed communities
> Communities of floating-leaf plants
>
> The assessment as raised by one point in each case in the event of the presence of: stretches of water; hollow trees of soft wood; walls, wall remains, broken stone or rubble; very varied habitat mosaic, pronounced contours; related woods of considerable age; and given a function as connecting biotope 1–3 points, depending on the quality.

The rationale underlying Wittig and Schreiber's method of evaluation was to provide a quick method with reference to planning, thus allowing little time for ecological surveys. They therefore proposed a survey of vegetation structure as an indication of habitat diversity (also used in some other methods of ecological evaluation, see p. 120). Would there, however, be a danger of overlooking areas of particular importance to rare or protected species? That question could equally well be applied to many other evaluation schemes and therefore Wittig and Schreiber's time-saving method would appear to be at least a useful preliminary method for identifying areas of nature conservation interest. However, it can not be denied that successful protection and management of urban habitats needs to be based on detailed surveys such as the comprehensive biotope mapping suggested by Sukopp and Weiler (1988) for German cities (p. 58).

5.6.3 Monitoring change in values of urban areas for nature conservation

Urban wildlife habitats are continually affected by new developments and are in a constantly changing environment. At the same time, perceptions of wildlife value in cities and towns seem to be changing from an appreciation of human-made habitats (parks and recreation grounds) to more natural habitats (green ways, urban wildlife areas). The changing urban environment and the more recent changes in perceptions of what is a valuable urban area soon outdate the results of any individual ecological evaluation. There would seem therefore to be a need to undertake evaluations in urban environments at more frequent intervals than in more natural areas. Data from surveys have, therefore, to be collated, analysed, stored, reviewed and communicated as efficiently as possible. This in turn demands good survey methods because classification of habitats and communities must precede evaluations (see p. 53). The use of detailed classifications, good surveys and well-planned monitoring of values of urban areas for nature conservation can not but help to generate a lot of data. Large quantities of data could be used as an argument not to undertake monitoring but information technology is now seen to be a key to successful information collection, storage and communication. Whereas Wittig and Schreiber's evaluation method can be commended for its simplicity, the use of databases and GIS systems provides new opportunities for much more detailed baseline surveys and subsequent more detailed evaluations to be undertaken.

5.7 ECOLOGICAL EVALUATION OF AQUATIC AND RIPARIAN HABITATS

5.7.1 Values and uses of wetlands and rivers

Wetlands are temporarily or permanently water-logged areas and include estuaries, wet peatlands, mires, marshes, fens, swamps, floodplains, deltas, coastal lakes and lagoons. A classification of wetland types assembled for the IUCN Directory of Wetlands of International Importance (Carp, 1978) has the following main groups:

1. Open sea shallow waters including the intertidal zone;
2. Sea bays and straits, including lagoons;
3. Mouth of rivers, that is tidal estuaries and deltas;
4. Coasts, including small islets;
5. Rivers and flood plains.

Many of these kinds of wetlands are species rich and some are among the most productive of all ecosystems in the world. The ecological value of some wetlands would therefore be expected to be high. As well as ecological value, some wetlands have an amenity and recreational value. The value of wetlands may also be expressed in monetary and economic terms (p. 69), especially in those instances where the wetland is a source of food and functions as a buffer against storm damage and flooding.

Despite the different values of wetlands, whether in ecological, amenity or in economic terms, more have been damaged and lost than any other kind of ecosystem. Wetlands have been drained, filled in, contaminated with pollutants and grossly over-exploited as a sink for human waste. This loss and damage to wetlands continues today as a result of the many pressures on land, especially around lagoons and estuaries. Some countries have now realized that there is a need for strong legislation to protect these productive and useful ecosystems. In the USA for example, Federal laws have been enacted for the protection of wetlands. A major chapter in the history of wetlands was the Ramsar Convention (p. 24) and today many countries have wetlands which are protected as a result of that Convention. The IUCN Wetland programme (IUCN, 1986) approved in 1985 has also prompted many initiatives including a network of experts to help in the management and conservation of wetlands.

Rivers and riparian habitats are valued for other reasons, in addition to their amenity and ecological importance. For example, much has been written about river corridors, meaning that waterways and riverbanks can facilitate dispersal and dispersion of wildlife (Spellerberg and Gaywood, 1993). However, like wetlands, running water systems have been subject to exploitation and varied impacts from a wide range of developments. In view of the damage to wetlands and waterways and the many pressures on these aquatic systems it is perhaps not surprising that there has been much interest in management and monitoring of wetlands (Brooks *et al.* 1989; Harris *et al.*, 1987) and also interest in evaluation of wetlands (Adamus, 1983; Leitch, 1989; Bardecki *et al.*, 1989).

Criteria for the selection of wetlands of international importance have been developed by a number of projects cosponsored by the IUCN and those criteria have been developed and tested (Box 5.3). As well as a global approach to evaluation of wetlands, different groups of wetlands have been the subject of different evaluation methodologies. Here we look briefly at methods of evaluation of standing water (such as marshes, lakes and reservoirs), rivers and riparian habitats. Most of the methods are directed at identifying the conservation value and are partly based on ecological variables.

5.7.2 Evaluation of marshes and standing water

In one of the most comprehensive surveys and evaluations of a large series of wetland habitats ever undertaken, Morgan and Boy (1982) came to the conclusion that there is no justification for using a scoring system to evaluate the conservation value of standing water. Instead, Morgan and Boy (1982) in their work in the Mediterranean region devised a rapid survey and classification system for the inland waters and then later assessed the conservation value of the sites (Morgan, 1982a, 1982b). The classification of the inland waters of the region was based on a wide range of physical, botanical and zoological characters then analysed by way of cluster analysis and ordination. The criteria used to assess the conservation value was particularly interesting because it was based on several, well-discussed criteria used for identifying potential nature reserves (p. 186). The following criteria were used:

1. Effectiveness as a conservation unit; Area, fragility, threats;
2. Representativeness;
3. Species richness, habitat diversity, abundance of waterfowl (see p. 141);
4. Naturalness;
5. Species rarity and site rarity;
6. Potential value;
7. Potential research use;
8. Potential educational and amenity use.

Although a scoring system was considered by Morgan and Boy (1982) they used it as a guide rather than a method of assessment because the importance of the different variables varies from site to site.

As with the ecological evaluation of other kinds of ecosystems, the choice of variables and the process to be used in the evaluation system seem to have been a major focal point of many evaluation attempts. Not surprisingly as the list of variables grows, so does the confusion as to which variables are most meaningful in terms of ecological value. This became all too clear when Friday (1988) attempted to undertake a conservation (and amenity) evaluation of some ponds in southern England. Four criteria were used (namely species richness, occurrence of species with a high conservation value, contribution to habitat diversity in the local area, amenity value) but because of the wide variation in habitats and biota it was concluded that all types of ponds have their own particular conservation value. This being the case, it seemed better not to score any particular pond but to consider the ponds as an integral part of the mosaic of habitats.

The use of ecological criteria seems to have had limited use in assessments of the nature conservation value of standing wetlands. This is partly because so many standing wetlands have an amenity, economic as well as a nature conservation value and therefore there is a belief that ecological criteria should be used together with other criteria. The perception of what is potentially valuable can however be misleading, especially where wetland sites are surrounded by varied land uses. In one of the most extensive investigations of the use of ecological data yet used in assessing the value of salt marshes, Oviatt *et al.* (1977) clearly showed that even when a marsh is surrounded by land which is intensively used, its ecological value is not necessarily diminished. A total of 17 biological parameters as well as morphological and geographical parameters were used by Oviatt *et al.* (1977) to evaluate 10 marshes in and around Rhode Island Sound (Providence River, MA). The biological parameters included standing crop, height and density of *Spartina*, abundance and species composition of fish, crustaceans and insects, and species richness and composition of birds. This intensive piece of work led the authors to conclude that marshes serve a number of functions and that there is little, if any, correlation between visual aesthetic perceptions of the marsh and its ecological characteristics. It would be interesting to investigate whether or not this conclusion holds true for other ecosystems and habitats.

Instead of using many ecological parameters, perhaps it would be more effective to use an indicator species to assess the conservation value of a wetland area. Birds have often been said to be good indicators of the state of the environment and even the dawn chorus could be said to be a useful indicator; any reduction in the dawn chorus over a number of years could be caused by deteriorating environmental conditions (Spellerberg, 1991). The use of birds and diversity of birds has been used in a particularly interesting manner by Harris *et al.* (1983) in and around the marshes of Green Bay, Lake Michigan, USA (Figure 5.5) as a basis for calculating ecological quality.

The primary objective at Green Bay was to test the usefulness of two indices of habitat diversity in assessing bird species diversity, that is edge diversity (Patton, 1975; Ghiselin, 1977) and vegetation cover type diversity (Milligan, 1981). Breeding bird censuses were made in permanent transects and a bird species diversity index was calculated using the Shannon-Wiener index (p. 47). The index of vegetation cover diversity was also calculated using the Shannon-Wiener index.

A shore development index or measure of shoreline irregularity as used by limnologists can equally well be used as a measure of edge diversity. That is:

$$ED = \frac{TE}{2\sqrt{(A\pi)}}$$

where *ED* is the edge diversity index, *TE* is the total linear edge, *A* is the area of the study plot and π is of course a symbol of the ratio of circumference to diameter (3.14). That is, if a lake had a shoreline in a perfect circle then the index would be 1. As the value of this ratio increases, there would be greater irregularity. In the study by Harris *et al.* the edge was taken as a measurement of interfaces between types of cover and between cover and water. The relationships between cover type diversity, edge diversity and bird species diversity were calculated via linear regression and they found that the equation that suggested the best assessment of ecological quality was the correlation of bird species diversity (dependent variable) with cover type diversity and edge diversity (independent variables). That is, it seems that in this study of freshwater coastal marshes, both the mosaic of vegetation and the diversity of vegetation was related to bird species diversity. A seemingly simple way to estimate habitat quality would therefore be to calculate an index of bird species diversity. Such indices can also be very useful and cost-effective in monitoring the ecological quality over time.

5.7.3 River systems and riparian habitats

Monitoring river water quality has long been the concern of water authorities and research organizations and consequently much research has been directed at both chemical and biological methods of monitoring. But monitoring requires baseline data and these are dependent on good classification systems. Indeed there is a close link between river classification systems and methods for evaluating the conservation value of rivers and riparian habitats. In Britain (HMSO, 1985; Wright *et al.*, 1989) Australia (MacMillan, 1984), South Africa (O'Keeffe *et al.*, 1987), and many other countries, river classification schemes and monitoring programmes have been instigated (see RIVPACS, p. 133). Despite the interest in classification there seems to have been less of an interest in methods for evaluating the conservation value of river systems and riparian habitats. There are, however two recent and particularly interesting systems which may well initiate more concern about evaluation for conservation.

The computer-based system devised by O'Keeffe *et al.* (1987) for classification and for assessment of the conservation value of rivers is of particular interest because it is cost-effective and lends itself to good communication between scientists and policy makers. The original aims of the River Conservation System (RCS) devised by O'Keeffe *et al.* (1987) were to identify those attributes of rivers which are important for their conservation, and to quantify the conservation value of a river. First, a list of attributes for the river, its catchment and biota were established using questionnaires

(a)

(b)

Figure 5.5 Aerial view of coastal marshes on Green Bay, Lake Michigan, USA. (a) was taken in April 1980 before 'greening' and the textural differences in cover type is accentuated. The water-vegetation interfaces are also more visible (note the opening created by muskrats – bomb crater effect). (b) was taken in August 1981. These photographs represent two years of a 15-year series used to assess vegetation changes in coastal marshes. Photographs kindly provided by Professor H.J. Harris.

and then weightings (+30 to −30) were applied. The final list of 41 attributes included the following: percentage of flow that is toxic effluent, importance as a migration route, length of river, habitat diversity, number of endemic fish species, importance of angling and a biotic index (see p. 53).

Scores for each attribute were calculated as a proportion of the weighting for the attribute. For example, 'endemic fish' has a weighting of 17 and any river with three or more endemic fish species would score 17. A river with only one endemic fish species would score 50% of the full weighting. The weighting represents the highest (or lowest) possible score for an attribute and the overall score for a river is calculated as the total for all attributes expressed as a percentage of the best possible scores (+ 17 in the case of 'endemic fish'). The computer program PROLOG is used to implement the RCS. A simplified example using a limited number of the attributes is shown in Table 5.7.

The evaluation of the conservation status of river corridor vegetation has been extensively studied by Slater *et al.* (1987) in Wales. A total of 105 habitat variables (31 river features, 42 bank features, 26 adjacent land features, 6 overall features) were recorded along entire lengths of each river. The slope of the bank was one of the variables recorded but interestingly enough, Learner *et al.* (1990) in a very thorough study of some river corridors in Wales found that bank slope was a poor predictor of the conservation status, at least in terms of the criteria they employed.

Vegetation data were collected intermittently along the rivers and from these data it was felt that three attributes (species richness, rarity, site uniqueness) could be of prime importance for assessing the conservation value of the plants in river corridors. Scores for species rarity (1–5) were allocated on the basis of national rarity (5) or local rarity (1). The score for site uniqueness was based on the frequency of occurrence with a river survey. The overall site class was calculated as follows:

$$\text{Site class} = \text{richness} + \text{rarity} + 0.5 \text{ uniqueness} / 2.5$$

Slater *et al.* (1987) looked at relationships between many of the habitat variables and vegetation data and concluded that of the attributes they chose, species richness and site uniqueness were more useful than species rarity.

5.8 CONCLUSIONS

Identification of species and classification of habitats are the foundations for any ecological evaluation. A common assumption running throughout the various methods of evaluation of conservation status has been that taxonomic competence would be available and could be confirmed. In only a few instances of description of ecological evaluations has there been

Table 5.7 Evaluation of river conservation status. A simplified example of the procedures for calculating the River Conservation Score based on only four of the many attributes. From O'Keeffe *et al.* (1987) with kind permission of the author

Attribute	Limits	Weighting
% sewage effluent	0%→>60%	−12
Indigenous fish spp.	0→>20	+ 9
Importance of angling	1 = none 4 = v. important	− 4
% natural vegetation	0%→>80%	+18

Using the Great Fish River (E. Cape Province) as an example system:

Attribute	Answers for Great Fish River	Score (Answer ÷ limit × weighting)
% sewage effluent	3%	$3 \div 60 \times -12 = -0.6$
Indigenous fish spp.	10 species	$10 \div 20 \times 9 = +4.5$
Importance of angling	2 = little importance	$2 \div 4 \times - 4 = -2$
% natural vegetation	30%	$30 \div 80 \times 18 = +6.8$

To calculate the overall score as a percentage of the best potential score:

Attribute	Potential best score	Actual score
% sewage effluent[a]	0(≡12)	−0.6(≡11.4)
Indigenous fish spp.	9	4.5
Importance of angling[a]	0(≡ 4)	−2 (≡2)
% natural vegetation	18	6.8
	43	24.7

Therefore the conservation status of the Great Fish River (on a scale of 0–100 is $24.7 \div 43 \times 100 = 57$

[a]To obtain comparable scores, negative attributes have first to be converted to an equivalent positive range.

concern expressed about the implications of lack of taxonomic expertise or mistakes arising from incorrect identification. With species richness, rarity, uniqueness and diversity being common variables in evaluations, the reliance on taxonomic competence is obvious. Unfortunately, the number of people with that competence has been decreasing and it is a problem which needs to be addressed urgently.

Evaluation of conservation status, whether by way of ecological indices only, or by a combination of subjective assessments, value judgements, quantifiable criteria and indices (such as aggregating the criteria), should never be accepted as being infallible. Although evaluations are designed to use biological and ecological parameters, allowances should be made for anomalies. For example, although high species richness is often thought to indicate high conservation value (old, diverse, unperturbed ecosystems with high species richness), some early successional stages (see succession in Appendix C) and certain ecosystems such as peatlands and heathlands are typically poor in plant species (but the few species of plants are almost certainly highly specialized, obligate inhabitants). In Britain, as opposed to Australia and New Zealand, there is a tendency to value mid-successional stages (plagioclimaxes: communities moulded by landuse and traditional management) rather than late successional stages or climax communities. In some instances, high species richness can arise from increased abundance of weeds arising from disturbance and the dumping of litter and rubbish. Those sites are not necessarily of high conservation status. One way of overcoming the problem is to consider only the number of indigenous species and exclude weed species in richness assessments.

The evaluation of species and communities can provide a basis with which to rank or establish priority rankings for the species or communities. To be able to do this seems to be central to the general aims of ecological evaluation for conservation but some fundamental problems have been suggested with regard to ranking systems. For example, Ehrenfeld (1976) in an appraisal of the consequences of assigning values to non-resources suggested that there are two fundamental problems with ranking systems, particularly with regard to the observation that ranking encourages choices when perhaps we should not be making choices. First, he drew attention to the not uncommon circumstance that evaluation may be based on incomplete knowledge of the species or community (see p. 37 for account of animal and plant distribution). It is impractical to undertake complete surveys of a site (encountering every species) and it is not always the case that the ecological requirements of a species under threat are fully documented. That being so, the disproportionate knowledge and the chances of overlooking some important feature does seem to make priority ranking somewhat presumptious. Second, he suggested that although ranking could be useful as an adjunct to decision making, priority ranking and the setting of natural area against natural area is unacceptable and totally unnecessary. In other words the need to conserve a particular community or species must be judged independently of the need to conserve anything else. Such an argument would have some difficulty in practice because resources are limited and it follows that there has to be a system whereby there is priority ranking at both a local and a national level.

Clearly there are many methods of ecological assessment and evaluation of biotic communities and only a few representative examples have been given here. It would almost seem that many workers interested in the use of ecological evaluation are not prepared to accept an existing scheme and therefore they propose a new scheme. The question is bound to arise, which evaluation scheme should I adopt and which is the most suitable? In view of the fact that there are so many unique circumstances, unusual conditions and differences in biogeography effects of mankind, it would be unrealistic to recommend any one evaluation scheme. In some countries, such as New Zealand and to a certain extent Australia, evaluations are being undertaken on natural areas and semi-natural areas. By way of contrast, cultural land-scapes, semi-natural habitats and plagioclimax communities at best are being evaluated in countries such as Britain and the USA. Therefore evaluation methods suitable for one country are not necessarily applicable to other countries. Does it matter therefore which method is used? I think it does, because clearly some methods of evaluation have serious limitations. In general, ecological evaluation should be based on sound biological concepts, the taxonomic and the mathematical limitations should be recognized and quantified. The methods should be appropriate for the region and the habitats being considered (for example methods for limestone pavements in Britain would not be appropriate for bird habitats in Sweden). Some methods seem to concentrate more on detailed mathematical and statistical calculations and so lose sight of the main objective. Simple methods seem more favourable than those which require large data sets and complex calculations.

In countries where little remains in the form of semi-natural areas, there is a need for modification of evaluation methods so that they can be used to evaluate potentially valuable areas: those areas which with some restoration of management could be improved. Such a category is used in the selection of nature reserves (section 6.6.2) but with a fragmented system of nature reserves, much could be made of the areas between reserves. Whether or not to have heathland restoration or the planting of new woodlands in an area could usefully be the subject of an ecological evaluation as could the evaluation of the potential value of an area.

Although all methods have their advantages and limitations, it would seem possible to adopt a fairly common conceptual plan for an ecological evaluation and one such plan is described in Chapter 8. Many of the methods of evaluation are based on habitats or on some other kind of ecological unit. It would seem important therefore that any ecological evaluation report should include not only specific details about sampling and recording methods (and limitations of those sampling methods, see p. 59) but also a list of the habitats which were recorded.

The use of 'ecological' indices seems attractive in theory but the methods have yet to be tested in case studies (see p. 163 for account of indices). At the same time it should be remembered that an aggregated index (multicriteria indices) is simply a piece of arithmetic derived from comparing and weighting the relative importance of criteria such as species richness, habitat diversity, vegetation complexity, uniqueness, rarity etc. It has been argued that these variables are not independent, that is species richness of some taxa is dependent on vegetation structure, and therefore it is unrealistic to compound the variables in one composite index.

On the other hand it could be argued that although they may be related, all the variables are qualities which should be conserved and so therefore aggregating them in one index may not be so unrealistic. Anselin *et al.*

(1989) have suggested one useful application of aggregated criteria by way of The Analytical Hierarchy Process.

An obvious question to ask if more than one criterion is used in an evaluation is: Do the criteria differ in their relative importance? If there is agreement about relative weights of criteria then perhaps this should be taken into consideration in the evaluation process. Margules and Usher (1984) have undertaken an intensive study of this particular aspect and found that with respect to criteria used for assessment of sites for nature reserves, there was considerable disagreement on the relative weights of criteria (p. 198)

The comments by Smith and Theberge (1987) and Bedward *et al.* (1991) about invalid procedures for adding or multiplying score classes for ordinal values (e.g. adding 1 for low, 2 for medium) are very timely. Common to many evaluations and assessments is the practice of adding score classes from ordinal values but clearly it is illogical to do this because the act of representing a range of measurements by a single figure is arbitrary and subjective.

Indices have a common use and are attractive in that they can neatly summarize a lot of information. However, it is particularly important that whenever an index is presented there is qualifying and supporting information such as:

1. A clear statement of how the index is calculated, preferably with a worked example;
2. The range of values that could be expected, i.e. the maximum and the minimum;
3. The variability and reliability, i.e. the range of values if several observers undertook independent measurements, and what would be the range of values if one observer carried out more than one calculation of the index at different times;
4. The method and calculation should be repeatable.

At its most simple, an evaluation could be based on a description of what is there in terms of habitats and taxonomic groups. That information should be qualified with lists of protected species, red data species or species unusually rare to that locality. The depth of such a descriptive report can be extremely varied as is discussed in Chapter 7. The following six points seems to emerge as being a suitable operational framework for an ecological evaluation.

1. The objective of the evaluation should be defined.
2. The criteria should be quantifiable rather than relying on subjective assessments.
3. The evaluation should be repeatable.
4. The evaluation should be based on biological principles.

5. The methods, results and analysis should be explained so that they can be understood by everyone who has an interest in the area being evaluated.
6. The costs in time and money need to be set against the depth and integrity of the ecological methods.

There seems to be general agreement that conceptual plans for an evaluation should include a statement about objectives and priorities (Roome, 1984). The method of ecological evaluation should then be adopted according to objectives and priorities. Pilot trials can be very valuable because they may reveal the weaknesses and limitations of the method. Also important is the need to ensure ecological validity (Margules and Usher, 1981, 1984) and sound mathematics (Dony and Denholm, 1985). Ecological criteria (such as rarity and representativeness) and other terms (such as sensitive species) should be quantified or defined. There seems to have been a temptation to develop methods which appear scientific but which in fact make use of scientific jargon and do not rely on sound scientific methods and biological principles. That is we should avoid pseudoscientific methods and rely more on simple but rigorous ecology that can be easily understood and provide a basis on which a consensus can be reached by ecologists and planners alike. Finally, there needs to be some thought about the costs and even more important a balance needs to be reached between costs of the evaluation and the depth of surveys underlying the evaluation. That is, the more detailed the survey, the greater the costs but it does not necessarily follow that more detailed surveys are based on better science.

REFERENCES

Ackery, P.R. and Vane-Wright, R.I. (1984) *Milkweed Butterflies*. British Museum (Natural History), London.

Adamus, P.R. (1983) A method for functional wetland assessment, Vol. 1, Critical review and evaluation concepts; Vol. 11, Assessment methods. Office of research, Development and technology, Federal Highway Administration, US Department of Transportation, Washington, DC.

Anselin, A., Meire, P.M. and Anselin, L. (1989) Multicriteria techniques in ecological evaluation: an example using the analytical hierarchy process. *Biological Conservation*, 49, 215–29.

Bardecki, M., Manning, E.W. and Bond, W. (1989) Practical methods for valuing wetlands: an assessment of three approaches, in the *Proceedings of the International Symposium on Wetlands and River Corridor Management* (eds J.A. Kusler and S. Daly), Berne, New York, USA, pp. 367–71.

Bedward, M., Pressey, R.L. and Nicholls, A.O. (1991) Scores and score classes for evaluating criteria: a comparison based on the cost of reserving all natural features. *Biological Conservation*, 56, 281–94.

Berry, P.M. (1983) The landscape, ecological and recreational evaluation of woodland. *Arboric. J.*, 7, 191–200.

Bezzel, E. and Reichholf, J. (1974) Die Diversitat als Kriterium zur Bewertung der Reichhaltigkeit von Wasservogel-Lebensraumen. *Journal of Ornithology*, 115, 50–61.

Brooks, R.P., Bellis, E.D., Keener, C.S., Croonquist, M.J. and Arnold, D.E. (1989) A methodology for biological monitoring of cumulative impacts on wetland, stream, and riparian components of watersheds, in the *Proceedings of an International Symposium on Wetlands and River Corridor Management* (eds J. A. Kusler, and S. Daly), Berne, New York, USA, pp. 387–98.

Brunner, M., Duhme, F., Muck, H., Patsch, J. and Wenisch, F. (1979) Kartierung erhaltenswerter Lebensraume in der Stadt. *Daz Gartenamt*, 28, 1–8.

Buckley, G.P. and Forbes, J.E. (1979) Ecological evaluation using biological habitats, an appraisal. *Landscape Planning*, 5, 263–80.

Carp, E. (1978) The IUCN directory of wetlands of international importance, in *Classification, Inventory, and Analysis of Fish and Wildlife Habitat. Proceedings of a National Symposium*. US Department of the Interior, Fish and Wildlife Service FWS/OBS–78/76, pp. 137–44.

Chanter, D.O. and Owen, D.F. (1976) Nature reserves: a customer satisfaction index. *Oikos*, 27, 165–7.

Disney, R.H.L. (1986) Assessments using invertebrates: posing the problem, in *Wildlife Conservation Evaluation* (ed. M.B. Usher) Chapman and Hall, London, pp. 271–93.

Domon, G., Bergeron, Y. and Mousseau, P. (1986) La hierarchisation des unites forestieres et des bois en milieu urbain sur la base de leur valeur ecologique. *Biological Conservation*, 37, 157–77.

Dony, J.G. and Denholm, I. (1985) Some quantitative methods of assessing the conservation value of ecologically similar sites. *Journal of Applied Ecology*, 22, 229–38.

Ehrenfeld, D.W. (1976) The conservation of non-resources. *American Scientist*, 64, 648–56.

Elton, C.S. (1966) *The Pattern of Animal Communities*. Methuen, London.

Elton, C.S. and Miller, R.S. (1954) The ecological survey of animal communities: with a practical system of classifying habitats by structural characters. *Journal of Ecology*, 42, 460–96.

Foster, G.N. (1991) Conserving insects of aquatic and wetland habitats, with special reference to beetles, in *The Conservation of Insects and Their Habitats* (eds N.M. Collins and J.A. Thomas) Academic Press, London, pp. 237–62.

Foster, G.N., Foster, A.P., Eyre, M.D. and Bilton, D.T. (1990) Classification of water beetle assemblages in arable fenland and ranking sites in relation to conservation value. *Freshwater Biology*, 22, 343–54.

Friday, L.E. (1988) The conservation and amenity value of ball-clay ponds in the Isle of Purbeck, Dorset, UK. *Biological Conservation*, 43, 165–80.

Fuller, R.J. (1980) A method for assessing the ornithological interest of sites for conservation. *Biological Conservation*, 17, 229–39.

Fuller, R.J. and Langslow, D.R. (1986) Ornithological evaluation for wildlife conservation, in *Wildlife Conservation Evaluation* (ed. M.B. Usher) Chapman & Hall, London, pp. 247–69.

Furse, M.T., Moss, D., Wright, J.F., Armitage, P.D. and Gunn, R.J.M. (1986) A *Practical Manual for the Classification and Prediction of Macroinvertebrate*

Communities in Running Water in Great Britain. Preliminary version. FBA, River Laboratory, East Stoke, Wareham.

Ghiselin, J. (1977) Analyzing ecotones to predict biotic productivity. *Environmental Management*, 1, 235–8.

Goldsmith, F.B. (1975) The evaluation of ecological resources in the countryside for conservation purposes. *Biological Conservation*, 8, 89–96.

Goodfellow, S. and Peterken, G.F. (1981) A method for survey and assessment of woodlands for nature conservation using maps and species lists: the example of Norfolk woodlands. *Biological Conservation*, 21, 177–95.

Götmark, F., Ahlund, M. and Eriksson, M.O.G. (1986) Are indices reliable for assessing conservation value of natural areas? An avian case study. *Biological Conservation*, 38, 55–73.

Harris, H.J., Milligan, M.S. and Fewless, G.A. (1983) Diversity: quantification and ecological evaluation in freshwater marshes. *Biological Conservation*, 27, 99–110.

Harris, H.J., Sager, P.E., Richman, S., Harris, V.A. and Yarbrough, C.J. (1987) Coupling ecosystem science with management: a Great Lakes perspective from Green Bay, Lake Michigan, USA. *Environmental Management*, 11, 619–25.

Helliwell, D.R. (1978) Survey and evaluation of wildlife on farmland in Britain: an 'indicator species' approach. *Biological Conservation*, 13, 63–73.

Helliwell, D.R. (1985) *Planning for Nature Conservation*, Packard Publishing, Chichester.

Hinton, R.F. (1978) Survey of ancient, semi-natural woodland in Hertfordshire, 1978. Unpublished report. Hitchin, Hertfordshire and Middlesex Trust for Nature Conservation.

HMSO (1985) Methods of biological sampling, a colonisation sampler for collecting macro-invertebrate indicators of water quality in lowland rivers. HMSO, London.

IUCN (1965) Project MAR. The conservation and management of temperate marshes, bogs and other wetlands. 2nd Vol. IUCN Publications, NS, No. 5.

IUCN (1986) *The IUCN Wetlands Programme 1985–1987*. IUCN Programme Series 5. IUCN, Gland, Switzerland.

Jarvinen, O. and Vaisanen, R. (1978) Habitat distribution and conservation of land bird populations in northern Norway. *Holarctic Ecology*, 1, 351–61.

Johnson, T.H. (1990) Avian endemism in the Philippines. *World Birdwatch*, 12, 6–7.

Kent, M. and Smart, N. (1981) A method for habitat assessment in agricultural landscapes. *Applied Geography*, 1, 9–30.

Klopatek, J.M., Kitchings, J.T., Olson, R.J., Kumar, K.O. and Mann, L.K. (1981) A hierarchical system for evaluating regional ecological resources. *Biological Conservation*, 20, 271–91.

Lancia, R.A., Miller, S.D., Adams, D.A. and Hazel, D.W. (1982) Validating habitat quality assessment: an example. *Transactions of the North American Wildlife and Natural Resources Conference*, 47, 96–110.

Learner, M.A,, Bowker, D.W. and Halewood, J. (1990) An assessment of bank slope as a predictor of conservation status in river corridors. *Biological Conservation*, 54, 1–13.

Leitch, J.A. (1989) Wetland economics and assessment: An annotated bibliography. Garland Publishing Inc., Hamden, Connecticut, USA.

Lloyd, C.S. (1984) A method for assessing the relative importance of seabird breeding colonies. *Biological Conservation*, **28**, 155–72.

Luff, M.L., Eyre, M.D. and Rushton, S.P. (1989) Classification and ordination of habitats of ground beetles (Coleoptera, Carabidae) in north-east England. *Journal of Biogeography*, **16**, 121–30.

MacMillan, L.A. (1984) A method for identifying small streams of high conservation status, in *Survey Methods for Nature Conservation* (eds K. Myers, C.R. Margules and I. Musto) South Australian Department of the Environment, pp. 319–42.

Margules, C. and Usher, M.B. (1981) Criteria used in assessing wildlife conservation potential: a review. *Biological Conservation*, **21**, 79–109.

Margules, C.R. and Usher, M.B. (1984) Conservation evaluation in practice. 1. Sites of different habitats in northeast Yorkshire, Great Britain. *Journal of Environmental Management*, **18**, 153–68.

Marsh, P. (1978) Formula for the needs of man and nature. *New Scientist*, **77**, 84–6.

Massey, M.E., Peterken, G.F. and Woods, R.G. (1977) Comparative assessments of woodlands for nature conservation. *NCC CST Notes*, **4**, NCC.

Milligan, M.S. (1981) *Resource Partitioning: Spatial and Behavioural Patterns in a Freshwater Coastal Marsh Avian Community*. MS Thesis, University of Wisconsin-Green Bay.

Morgan, N.C. (1982a) An ecological survey of standing waters in North West Africa: II. Site descriptions for Tunisia and Algeria. *Biological Conservation*, **24**, 83–113.

Morgan, N.C. (1982b) An ecological survey of standing waters in North West Africa: III. Site descriptions for Morocco. *Biological Conservation*, **24**, 161–82.

Morgan, N.C. and Boy, V. (1982) An ecological survey of standing waters in North West Africa: I. Rapid survey and classification. *Biological Conservation*, **24**, 5–44.

Nilsson, S.G. and Nilsson, I.N. (1976) Hur skal naturomraden varderas ? Exempel fran fagellivet i sydsvenska sjoar. *Fauna och Flora*, **4**, 136–44.

O'Connor, K.F., Overmars, F.B. and Ralston, M.M. (1990) *Land Evaluation for Nature Conservation*. Department of Conservation, Wellington, NZ.

O'Keeffe, J.H., Danilewitz, D.B. and Bradshaw, J.A. (1987) An 'expert system' approach to the assessment of the conservation status of rivers. *Biological Conservation*, **40**, 69–84.

Ogle, C.C. (1981) The ranking of wildlife habitats. *New Zealand Journal of Ecology*, **4**, 115–23.

Ogle, C.C. (1982) Wildlife and wildlife values of Northland. Fauna Survey Unit Report No. 30, Wellington, New Zealand Wildlife Service, Department of Internal Affairs.

Oviatt, C.A., Nixon, S.W. and Garber, J. (1977) Variation and evaluation of coastal salt marshes. *Environmental Management*, **1**, 201–11.

Patton, D.R. (1975) A diversity index for quantifying habitat 'edge'. *Wildlife Society Bulletin*, **3**, 171–3.

Peat, J.R. (1984) A partially objective method for the ecological evaluation of biological communities. PhD Thesis, The University of Southampton.

Perring, F.H. and Walters, S.M. (1962) *Atlas of the British Flora*. BSBI/Nelson, London.

Peterken, G.F. (1974) A method for assessing woodland flora for conservation using indicator species. *Biological Conservation*, 6, 239–45.

Peterken, G.F. (1977) Habitat conservation priorities in British and European woodlands. *Biological Conservation*, 11, 223–36.

Peterken, G.F. (1981) *Woodland Conservation and Management*. Chapman & Hall, London.

Peterken, G.F. (1991) Ecological issues in the management of woodland nature reserves, in *The Scientific Management of Temperate Communities for Conservation* (eds I.F. Spellerberg, F.B. Goldsmith and M.G. Morris), Blackwell Scientific Publications, Oxford.

Peterken, G.F. and Game, M. (1984) Historical factors affecting the number and distribution of vascular plant species in the woodlands of central Lincolnshire. *Journal of Ecology*, 72, 155–82.

Prater, A.J. and Lloyd, C.S. (1987) Birds, in *Biological Surveys of Estuaries and Coasts* (eds J.M. Baker and W.J. Wolff) Cambridge University Press, Cambridge.

Refseth, D. (1980) Ecological analyses of carabid communities – potential use in biological classification for nature conservation. *Biological Conservation*, 17, 131–41.

Roome, N.J. (1984) Evaluation in nature conservation decision-making. *Environmental Conservation*, 11, 247–52.

Saaty, T. (1977) A scaling method for priorities in hierarchical structures. *Journal of Mathematical Psychology*, 15, 234–81.

Saeijs, H.L.F. and Baptist, H.J.M. (1977) Wetland criteria and birds in a changing delta. *Biological Conservation*, 11, 251–66.

Savage, A.A. (1982) Use of water boatmen (Corixidae) in the classification of lakes. *Biological Conservation*, 23, 55–70.

Schamberger, M. and Krohn, W.B. (1982) Status of the habitat evaluation procedures. *Transactions of the North American Wildlife and Natural Resources Conference*, 47, 154–64.

Sinden, J.A. and Windsor, G.K. (1981) Estimating the value of wildlife for preservation: a comparison of approaches. *Journal of Environmental Management*, 12, 111–25.

Slater, F.M., Curry, P. and Chadwell, C. (1987) A practical approach to the evaluation of the conservation status of vegetation in river corridors in Wales. *Biological Conservation*, 40, 53–68.

Smith, P.G.R. and Theberge, J.B. (1987) Evaluating natural areas using multiple criteria: theory and practice. *Environmental Management*, 11, 447–60.

Smith, S.D. (1984) An evaluation method for urban open spaces and its relevance in urban nature conservation. Environmental Sciences undergraduate project, Southampton University.

Spellerberg, I.F. (1991) *Monitoring Ecological Change*. Cambridge University Press, Cambridge.

Spellerberg, I.F. and Gaywood, M.J. (1993) Linear Features: wildlife corridors and linear habitats. English Nature, Peterborough.

Stroud, D.A., Mudge, G.P. and Pienkowsk, M.W. (1990) Protecting internationally important bird sites. Nature Conservancy Council, Peterborough.

Sukopp, H. and Weiler, S. (1988) Biotope mapping and nature conservation strategies in urban areas of the Federal Republic of Germany. *Landscape and Urban Planning*, **15**, 39–58.

Sukopp, H., Kunick, W. and Schneider, C. (1980) Biotope mapping in the built-up areas of West Berlin. *Garten und Landschaft*, 7/80, 565–9.

Sukopp, H. (1990) *Urban Ecology*, SPB Academic Publishing, The Hague.

Tans, W. (1974) Priority ranking of biotic natural areas. *Michigan Botanist*, **13**, 31–9.

Thomas, W.A. (1972) *Indicators of Environmental Quality*. Plenum Press, New York.

Usher, M.B. and Edwards, M.E. (1986) The selection of conservation areas in Antarctica: an example using the arthropod fauna of Antarctic islands. *Environmental Conservation*, **13**, 115–22.

US Fish and Wildlife Service (1980) *Habitat Evaluation Procedures (HEP)*. Division of Ecological Services, US Fish and Wildlife Service, Department of the Interior, Washington, DC.

Vane-Wright, R.I., Humphries, C.J. and Williams, P.H. (1991) What to protect? Systematics and the agony of choice. *Biological Conservation*, **55**, 235–54.

van der Ploeg, S.W.F. and Vlijm, L. (1978) Ecological evaluation, nature conservation and land planning with particular reference to methods used in the Netherlands. *Biological Conservation*, **14**, 197–221.

Ward, S.D. and Evans, D.F. (1976) Conservation assessment of British Limestone pavements based on floristic criteria. *Biological Conservation*, **9**, 217–33.

Williams, (1964) *Patterns in the Balance of Nature*. Academic Press, New York.

Wittig, R. and Schreiber, K-F. (1983) A quick method for assessing the importance of open spaces in towns for urban nature conservation. *Biological Conservation*, **26**, 57–64.

Wright, J.F., Armitage, M.T., Furse, M.T. and Moss, D. (1989) Prediction of invertebrate communities using stream measurements. *Regulated Rivers: Research and Management*, **4**, 147–55.

6

Evaluation and assessment of areas for protection

6.1 INTRODUCTION

Very often economics, politics and imminent threats rather than ecology or conservation interests dictate what areas are selected and when. That is, financial stringencies may result in a second best site being selected, new sites may be designated at a time most suitable for political gain and unexpected developments or proposed changes in land use have sometimes forced a conservation organization to establish a protected area as a matter of urgency. Nevertheless there have been many discussions about the most appropriate criteria for selection of protected areas such as nature reserves and those discussions continue. For example, Götmark and Nilsson (1992) recently reviewed criteria for protection of natural areas in Sweden and concluded that scientific criteria (as opposed to recreational and landscape values) should have priority for conservation.

This chapter is concerned with five aspects of protected areas; a brief historical account, the objectives and categories of protected areas, those areas primarily allocated for the protection of landscapes, wilderness areas and finally those areas primarily allocated for conservation of nature. For convenience, 'nature reserve' is used in a broad sense and includes all areas set aside primarily for conservation of nature.

6.2 A HISTORICAL PERSPECTIVE

Although national parks are only one category of protected area, they have been a 'flagship' for the promotion of protected areas. Furthermore, the selection of protected areas, the discussions about objectives and the research into criteria on which to base selections is epitomized in the establishment and gradual increase in national parks throughout the world.

The establishment of protected areas, such as national parks, has been undertaken for many years. Early designations took place in 1864 and in 1872, when the Yosemite National Park and the Yellowstone National Park

(900 000 ha) were respectively established in North America. Today, the National Parks of North America include more than 116 000 square miles of parks and recreational areas. Interestingly, the criteria used for the establishment of the Yellowstone National Park were adopted 97 years later by the General Assembly of the IUCN at the 1969 meeting in New Delhi. The national parks of North America have been successfully selected using ecological criteria and they now provide a network of representative ecosystems, biomes and habitats (Curry-Lindahl, 1988). Some of the national parks and national forests (first set aside in the USA in 1891) are wilderness areas and as such provide for both conservation of biodiversity and recreation. The process of designation continues and in 1980, the Alaskan Wilderness Area of 160 000 square miles received protection.

Elsewhere, Australia and New Zealand established national parks as long ago as 1886 and 1894 respectively. In Europe, the first national parks were established by Sweden (1909), Russia (1912), Switzerland (1914) and Spain (1918). In Britain, The National Parks and Access to the Countryside Act 1949 was an important step towards the designation of national parks which were to be extensive areas of beautiful and wild country, consisting largely of mountains and moors (with associated farm lands), heaths, rocky and infertile coasts. Ten National Parks were designated by the National Parks Commission (note the changes in status of these organizations as shown in Box 6.1): Peak District, Lake District, Snowdonia, Dartmoor, Pembrokeshire Coast, North York Moors, Yorkshire Dales, Exmoor, Northumberland and the Brecon Beacons). These ten National Parks are in England and Wales but none occur in Scotland. There are, however, 40 National Scenic Areas in Scotland; regions identified by the Countryside Commission for Scotland as being of outstanding scenic beauty.

6.3 PROTECTED AREAS: OBJECTIVES, CATEGORIES AND AREA

6.2.1 Objectives and categories

The UNESCO Man and the Biosphere Programme of Biosphere Reserves has a worldwide basis but there are several other international categories of protected areas. Some of these are listed in Table 6.1, together with their objectives. There has to be a variety of protected areas to allow for a range of suitable management objectives whether it be selective forestry, controlled grazing, a combination of recreation and conservation or total protection of the wildlife (Mackinnon *et al.*, 1986). The variety of protected areas on a countrywide basis can also be extensive (Tables 6.2, 6.3).

Box 6.1 Changes in status of government 'conservation organizations' in Britain.

National Parks and Access to Countryside Act 1949

Formation of the National Parks Commission.
The Nature Conservancy is Established.

Countryside Act 1968

Countryside Commission replaces the National Parks Commission.

The Nature Conservancy Act 1973

The Nature Conservancy is split into two bodies: the Nature Conservancy Council (responsible to the Secretary of State for conservation and responsible for management of nature reserves); the Institute of Terrestrial Ecology (a research organization remaining in the Natural Environment Research Council).

The Environmental Protection Act 1990

Reorganization of the Nature Conservancy Council:

1. The Nature Conservancy Council for England (English Nature). In England the NCC and Countryside Commission remain separate.
2. Countryside Council for Wales (Country Commission and NCC in Wales combined).
3. Nature Conservancy Council for Scotland – in 1992 to combine with the Countryside Commission for Scotland to become Scottish Natural Heritage.
4. The Joint Nature Conservation Committee (coordinating role and responsible for European and international matters).

Maintaining a database on protected areas is one of the important activities of the WCMC and from time to time the WCMC publishes data showing the growth in number of protected areas (Figure 6.1).

It is not unusual for a protected area to be established with several objectives or functions in mind. For example in the UK, National Nature Reserves are established with five functions in mind: conservation of species and habitats, research, demonstration and advice, education, amenity and access. Indeed this multifunctional role has very important implications in a world where nature reserves could by reasons of their protective role be at conflict with the needs of local communities. It is for this reason that the WWF in a document on the conservation of tropical forests and biodiversity has advocated that tropical protected areas should be conceived as elements of overall schemes for integrated rural development (WWF, 1991).

The WWF has been particularly active in promoting the value of protected areas. For example, it was instrumental in promoting the idea of an EEC Directive on the Protection of Natural and Semi-natural Habitats and

Table 6.1 The range of protected areas throughout the world and the objectives. From Reid and Miller (1989) with permission of the World Resources Institute

Conservation Objective	Scientific Reserve (1)	National Park (2)	Natural Monument Reserve (3)	Natural Wildlife Reserve (4)	Protected Landscape Reserve (5)	Resource Reserve (6)	Anthropological Reserve (7)	Multiple-Use Area (8)	Extractive Reserve	Game Ranch Area	Recreation Area
Maintain sample ecosystems in natural state	+	+	+	+	o	o	+				
Maintain ecological diversity and environment regulation	o	+	+	+	o	o	+	o	o	o	o
Conserve genetic resources	+	+	+	o	o	o	+	o	o	o	o
Provide education, research and environmental monitoring	+	+	+	+	o	o	+	o	o	o	o
Conserve watershed, flood control	+	o	+	+	o	o	o	o	o	o	o
Control erosion and sedimentation	o	+	o	o	o	o	o	o	o	o	o
Maintain indigenous uses or habitation					+	o	+	o	o	o	o
Produce protein from wildlife				o	o	o	o	+	+	+	
Produce timber, forage or extractive commodities					o	o	o	+	+		

Protected area designation (IUCN category number)

Table 6.1 continued

Provide recreation and tourism services	+	+	+	o	+	o	+	o	+
Project sites and objects of cultural, historical, or archaeological heritage	o	o	o	+	+	o	o	o	o
Protect scenic beauty	o	+	+	+	+	+	+	o	o
Maintain open options, management flexibility, multiple-use				o	+	+	+		
Contribute to rural development	o	+	o	+	o	o	+	+	o

Table 6.2 Number and extent of protected areas in New Zealand (1987)

Protected Area Class	No.	Area (ha)	% NZ area
National Park (+ Paparoa National Park, 30000 ha)	(12)	(2 203 317)	(8.20)
Conservation Park (Formerly State Forest Park)	19	1 754 702	6.53
Scenic Reserve	1242	340 497	1.27
Ecological Area	92	219 283	0.82
National Park Specially Protected Area	5	190 234	0.71
Nature Reserve	50	185 577	0.69
Wildlife Refuge	131	74 000	0.28
Recreation Reserve	1710	53 242	0.20
Sanctuary Area	14	16 288	0.06
Wildlife Management Reserve	146	12 484	0.05
Scientific Reserve	49	9 249	0.03
Open Space Covenant	178	5 000	0.02
Protected Private Land	64	3 209	0.01
Areas of Historic Interest	162	2 739	0.01
Marine Reserve	2	2 687	0.01
Conservation Covenant	19	589	–
Wildlife Sanctuary	16	424	–
Total (National Total Area (excluding island territories) = 26 868 400 ha)	(3911)	(507 352)	(18.88)

Data kindly provided by D.A. Norton.

of Wild Fauna and Flora. The fundamental purpose of the Directive is to establish a comprehensive network of protected areas (Special Areas of Conservation: 'Natura 2000') by the year 2000. There are positive indications that this Directive will soon be agreed by Member States.

6.2.2 Area

The number of protected areas on a worldwide basis and the total protected area is not easy to monitor because of inaccuracies in measurement and because of poor communication. The WCMC figures show that the number of protected areas increased quite sharply from the early 1970s onwards. In 1988 McNeely estimated that about 4% of the world's land surface has been designated as protected areas. That sounds encouraging but different areas receive different levels of protection, some in name only.

It is difficult to choose any particular country on the basis of its success in designating protected areas for the conservation of biodiversity and landscapes. Some of the smaller countries (under a million km²) are however of interest. For example, Venezuela (912 050 km²) has a particularly large area of protected land. More than 44% is protected by Special Admin-

Table 6.3 Number and extend of protected areas in Britain (1992). Data kindly supplied by Brian Davies, NCC

Status	Number	Total area	Correct to	% GB area
Biosphere Reserves	13	35384 ha	31/12/92	0.15
Local Nature Reserves	290	12954 ha	30/9/92	0.056
Marine Nature Reserves	2		31/3/91	
National Nature Reserves	242	168107 ha	31/3/91	0.73
Non-governmental organization:				
RSPB reserves	107	70614 ha	16/1/89	0.30
RSNC reserves	1673	54745 ha	16/1/89	0.23
RAMSAR	56	179744 ha	31/12/92	0.78
Sites of Special Scientific Interest (SSSI)	5671	1778474 ha	31/3/91	7.73
Special Protection Areas	61	150992 ha	31/12/92	0.65
Areas of Outstanding Natural Beauty	38	19324 sq km	31/3/90	8.4
Heritage Coasts	43	1460 km in length	31/3/90	
National Parks	10	13648 sq km	31/3/90	5.93
National Scenic Area	40	1001800 ha	31/3/90	4.35
National Trust[a]	350	56000 ha	31/3/85	0.24

[a]The National Trust figures are for SSSIs in their ownership.

istration Regimes (ABRAES) and new protected areas continue to be designated. About 20% of Venezuela's protected land consists of a particularly important biodiversity area, the Amazon Federal territory (TEA). An area authority, Sada-Amazonas (Appendix B) administers the management plan for the Territory.

In yet smaller countries such as New Zealand (268 676 km^2) and the UK (244 755 km^2) there are comparatively large areas designated for the protection of nature and landscapes (Tables 6.2 and 6.3). New Zealand with a population of 3.5 million (1990) and a population density of 12.3 per km^2 has a variety of protected areas which amount to about 19% of the area of New Zealand. In Britain with a population of 55 million (population density of about 230 per km^2) it is perhaps surprising to find that there is a similar amount (20%) of land designated as protected areas but that includes national parks and other landscape conservation areas (Table 6.3). Excluding National Parks, National Scenic Areas and National Trust reserves, only 9.6% of land is designated as protected areas.

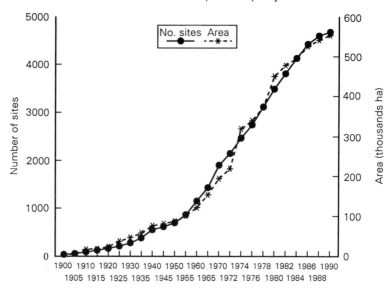

Figure 6.1 The growth in the number of nationally protected areas (nature reserves, national parks, natural monuments and wildlife sanctuaries). Data kindly supplied in 1990 by the World Conservation Monitoring Centre and redrawn after Speller-berg (1991).

6.4 LANDSCAPE EVALUATION

6.4.1 Protection of landscapes

Whereas nature reserves are designed primarily to conserve wildlife (but see p. 93 for additional uses), many other kinds of protected areas are designated primarily to conserve landscapes or landscape features of histori-cal interest. Examples include National Parks, National Monuments, Areas of Outstanding Natural Beauty (AONB) and Areas of Great Landscape value. Protected Landscapes could also be included here but they are differ-ent in that the Protected landscape provides a clear and legitimate alterna-tive to the National Park: protected landscapes are designated to maintain nationally significant natural landscapes which are characteristic of har-monious interactions of people and land but at the same time provide opportunities for public enjoyment (Lucas, 1992).

As with protected areas designated for nature conservation, various selec-tion procedures and criteria have been developed for the selection of pro-tected landscapes. The methods used for landscape evaluation and the philosophy underlying landscape evaluation have been researched for at least three decades, mainly in Europe and North America, by geographers, policy makers and planners (Dearden and Sadler, 1989).

The objectives of landscape evaluation do not seem to be without question but there can be problems. One problem is that different people like different landscapes (likes are sometimes affected by environmental and socioeconomic backgrounds; Tips and Savasdisara, 1986a). As there may be preference differences, is it better to try and obtain a consensus among people using the area or leave the evaluation to a few planners and policy makers? If you ask for preferences do you ask people to comment on the landscape as a whole or do you ask for comments on particular features? In the early 1970s, for example, the basis of landscape preference studies was to consider landscapes in their totality rather than base evaluations on a number of landscape features and in particular to take into consideration the public's perception of landscape quality (Penning-Roswell, 1981).

6.4.2 Landscape evaluation methods and practice

In an historical review (1960–1970) of landscape evaluation Penning-Rowsell (1981) identified three different approaches which appeared chronologically as intuitive methods, statistical methods, and an emphasis on landscape preferences. In North America the intuitive method was a broad approach used in different ways by different authorities to identify those areas of land which were desirable, informal recreational areas such as wilderness areas and forest parks. This broad approach varied according to the policies of the authorities and was based largely on an intuitive classification of landscapes based on visual qualities and surveys of the area under consideration. However, field surveys may require several months' work and so can be costly. Not surprisingly, there have been attempts to develop methods to replace field surveys. For example, a regression model has been developed for landscape preferences based on measurements from photographs as independent variables (Shafer *et al.*, 1969).

Much work has gone into quantifying those elements which contributed to landscape quality and which could be achieved either in field surveys or less expensively by using surrogate material such as maps, photographs or computer simulations of the area. The main objective has been to quantify the data as a prelude to various kinds of statistical analysis. For example, in the USA, some of the early techniques by Leopold (1969) concentrated on the criteria which can be used to judge and rank landscapes. The physical, biological and human interest features of the landscape (sometimes as many as 46 features) would be identified and scored either by using empirical measurements or value judgements. Other attempts at objectivity have been directed at investigating the correlation between perceived attractive landscape and determinants of attractive landscape such as uniqueness and diversity. Ribe (1986), for example, found that there is a good correlation between diversity of landscape features (based on the Shannon-Weiner diversity Index, p. 47) and perceived scenic beauty.

Landscape evaluation relies on land classification (see p. 53) as the following example demonstrates. The Linton (1968) method for assessment of scenery as a natural resource which evolved from previous work (Fines, 1968) was developed in Scotland (Gilg, 1974, 1976). The method was based on the simple parameters of 'landform landscapes' and land use. Linton proposed six categories of landform landscapes as follows:

1. Lowland; land below 500 ft;
2. Hill country; relative relief of less than 1000 ft;
3. Bold hills; usually in excess of 1200 ft;
4. Mountains;
5. Plateau uplands; 'There are several upland areas, of which the Mona-dhliath are the loftiest and most extensive examples, which have hilly margins, sometimes with decidedly bold relief, but whose interior portions are not visible except from higher ground. They offer extensive areas of apparently rolling relief of generally 300 ft or less, though they may contain some deeply hidden valleys';
6. Low uplands. 'Many areas in Scotland lie at altitudes below 1000 ft and have no more relative relief than true lowlands but certainly cannot be classed with them'.

Categories 2, 3 and 4 constitute an ascending sequence in which the forms are at least interesting and it is suggested that they are given arbitrary ratings of 5, 6 and 8 points (Table 6.4). Low uplands are rated 2 and plateau uplands 3 points. Bonus points (weighting) could be given for water: 2 points 'to areas with water in the foreground and middle distance of the view, and 1 for the uplands with a significant water element in view.'

The land use landscapes used by Linton were as follows:

1. Urbanized and industrialized landscapes;
2. Continuous forests;
3. Treeless farmland;
4. Moors (elevated regions of wet acid peat, typically dominated by ericoid shrubs such as heathers);
5. Varied forest and moorland landscapes;
6. Richly varied farming landscapes;
7. Wild landscapes.

Urbanized and industrialized landscapes score negative points and as much as −6 or −7 was suggested by Linton. Continuous forest also scores negative points and the suggested value was −2. Only 1 point is awarded to treeless farmland. Moors score 3, richly varied landscapes score 5 and wild landscapes 6. The two parameters of landform and land-use are scored

Table 6.4 A method for evaluation of 'scenery' as a natural resource. This method is based on the two simple criteria of landform (column) and land use (row) with scores allocated on a subjective basis. From Linton (1968) with kind permission of the Royal Scottish Geographical Society.

Category and rating	Urbanized and industrialized −5	Continuous forest −2	Treeless farmland +1	Moorland +3	Varied forest and moorland +4	Richly varied farmland +5	Wild landscapes +6
Lowland 0	Ayrshire Coast −5	Culbin Forest −2	Machars of Wigtown +1	Moors of Wigtown +3	Cromar +4	Strathearn +5	Colonsay +6
Low uplands 2	N.E. Lanarkshire −3	Darnaway Forest 0	N.E. Berwickshire +3	Loch Shin +5	Methven and Glenalmond +6	Gifford (East Lothian) +7	Gruinard Bay +8
Plateau uplands 3	(No example) −2	Kielder Forest +1	Falahill (Midlothian) +4	Monadhliath +6	Potrail Water +7	(No example) +8	Mid Argyll +9
Hill country 5	Bathgate Hills 0	Nethy Forest +3	Gala Valley +6	Crawick Water +8	Speyside +9	Middle Tweed +10	Loch Moidart +11
Bold hills 6	Leadhills – Wanlockhead +1	Strathyre Forest +4	Loch Chon +7	Dalveen Pass +9	Bennachie +10	(No example) +11	Morvern and Sunart +12
Mountains 8	Kinlochleven +3	Loch Lubnaig +6	Glen Elchiag +9	Glen Clova +11	Achnashellach +12	(No example) +13	Coigach +14

separately and the landscape score is derived as the sum of the scores of those two components (Table 6.4).

In a generally favourable assessment of Linton's method in Scotland, Gilg (1974, 1976) drew attention to the need for a clear statement of the exact scale on which to base the evaluation and also a need for a clear direction with regard to the spatial frequency of sampling. In considering scale and frequency of sampling, Gilg used 10 km squares divided into 2 km squares. Data could then be collected for each km square along rows and columns for up to 20 km. One of the difficulties raised by Gilg is the difficulty of interpreting between two or more land uses or land forms when they both occur within the same square. A solution suggested by Gilg is simply to apply a *pro rata* apportionment when the percentage cover of each land use falls below 60%.

6.5 WILDERNESS AREAS

There seem to be few definitions of a wilderness area yet they are the subject of many reports, books and papers. In referring to wilderness areas, the more common words used are remoteness, unmodified and primitive. Perhaps the best description of wilderness areas can be found in the The Wilderness Act of 1964 (USA), that is an area of land which retains its primitive features, which does not have permanent improvements or human habitation and which is protected and managed to preserve its natural conditions (see p. 87). Two very concise definitions of wilderness areas are: areas 'big enough to absorb a two weeks' pack trip' (Leopold, 1921); 'an undeveloped land still primarily shaped by the forces of nature' (McCloskey and Spalding, 1989).

Examples of wilderness areas may be found among national parks, forests, reserves, environmentally significant areas (ESAs; not to be confused with the environmentally sensitive areas established in Britain under the Agricultural Act 1986) and large expanses of mixtures of designated and undesignated regions. In the first reconnaissance and inventory of wilderness on the land areas of the world, McCloskey and Spalding (1989) concluded that one-third of the earth is still wilderness. They based their inventory on areas over 400 000 ha but 41% of the area included the polar regions. Almost certainly the largest and most unique wilderness area is Antarctica and for that reason there have been many suggestions that Antarctica be designated a world park, indeed the world's first international park.

That Antarctica should become a world park seems reasonable on the grounds that it is a remote and relatively undisturbed area. But as wilderness areas become ever more scarce, are there opportunities to identify, evaluate and select wilderness areas in countries such as Australia, America and the USSR? In Australia, the wilderness continuum concept has been developed which allows a precise assessment of wilderness resources and which identifies those attributes which contribute to wilderness quality (Lesslie and Taylor, 1985, Lesslie, 1991).

The computer-based methodology developed by Lesslie *et al.* (1988a) is designed to measure variation in naturalness and remoteness across the landscape using prescribed environmental indicators. There are three major steps in the process: establishment of primary databases, derivation of indicator values, analysis of indicator values to evaluate wilderness quality. The four indicators used in this procedure are remoteness (distance from settlement, remoteness from access, aesthetic naturalness and biophysical naturalness. Numeric values were calculated for each of these indicators on a grid of many sampling points using GIS. For example, within the State of Victoria, Australia (227 620 km^2 in area), a total of 3.310^5 grid points were used. An estimation of total wilderness quality may be simply derived by standardizing and combining the four wilderness indicator values. Information about wilderness resources can usefully be provided by these techniques but Lesslie (1991) concluded that no objective measure of wilderness quality can be used to decide whether or not areas should be set aside for protection. Such decisions, he felt, are inevitably judgemental, requiring comparative assessments of the social worth of alternative and usually conflicting land-use opportunities.

The recent advances in GIS (with powerful methods for spatial data analysis) will no doubt contribute to further development of this computer-based method of wilderness evaluation and make it a very important tool for management and monitoring change as well as for evaluating vast expanses of wilderness areas. (An Australian National Wilderness Inventory is currently being conducted by the Australian Federal Government using GIS and Lesslie's methods (Lesslie *et al.*, 1988b).) As well as inventories of wilderness areas, there are now well-established methods for evaluating the biotic diversity of wilderness areas such as ESAs. For example, Smith and Theberge (1986a) developed a method for evaluating biodiversity of ESAs in the northwest territories of Canada based on five ecological criteria including species richness of conspicuous vertebrate species and vascular plants.

Management of protected areas should be based on good ecological science but in reality many other factors have to be considered, so many that an interdisciplinary framework has to be adopted. An interdisciplinary framework was clearly demonstrated in the concept of The Limits of Acceptable Change (LAC). Although not a new concept, it was developed by Stankey *et al.* (1985) as part of an innovative and continuing effort to try and improve wilderness recreation management in the face of carrying capacity of wilderness areas being exceeded. Stankey *et al.* (1986) proposed a method in which the amount of change to be accepted was defined in explicit terms, and was based on quantitative standards. That approach identified management actions which were necessary to prevent further changes and which also provided a sound basis for monitoring and evaluating management performance. Understandably, the LAC concept has to be based on perceived wilderness conditions and so therefore evaluation can take place in one of two stages: the evaluation as a basis for designation, evaluation of the effects of management. Nine steps were suggested by Stankey *et al.* (1986) as follows:

1. Identification of area concerns, issues and legal constraints.
2. Opportunity classes are defined and described. These are hypothetical descriptions of the range of conditions that are considered likely to be maintained or restored. For example, primitive, motorized, rural and urban: primitive and semi-primitive non-motorized classes apply to wilderness areas.
3. Indicators of resource and social conditions are selected. Resource factors could range from trail conditions to the status of wildlife populations and social factors could range from campsite solitude to conflicts regarding party size.
4. An inventory is made of the existing resource and social conditions which in turn provides a basis for mapping and analysis.
5. Standards are specified for resource and social indicators for each of the opportunity classes.
6. Alternative opportunity classes reflecting area issues and concerns and also existing resource and social conditions are identified.
7. Management actions for each alternative are identified.
8. Preferred alternatives are evaluated and assessed.
9. Actions are implemented and monitored.

The Wilderness Act 1964 (USA) refers to 'preservation of wilderness character' but also goes on to say that 'wilderness areas shall be devoted to the public purposes of recreational, scenic, scientific, educational, conservation and historic use'. Clearly, there needs to be careful planning and management so as to avoid as much conflict of use as possible. The LAC system for wilderness planning is one very good system on which to base wilderness planning and is one which can help to avoid such conflicts. The concept of LAC seems to have a range of applications in a wide setting, and would seem to go beyond a use in connection with only wilderness areas. Stankey and others are now looking at the application of LAC in coastal and marine settings as well as in non-recreational locations (G.H. Stankey personal communication).

6.6 NATURE RESERVES: CRITERIA AND SELECTION

6.6.1 The variety of criteria

Different sets of criteria are used for the selection of different categories of protected areas. There are, however, some criteria in common use which include number of habitats and/or species, naturalness and rarity of both habitats and species and area (Usher, 1980, 1986b). Other commonly used criteria include threat of human interference, amenity value, educational value, scientific value, recorded history, typicalness and representativeness (these are defined in the following pages). Only a few of these may be appropriate for some regions. In the Antarctic for example, only two criteria (uniqueness and outstanding scientific interest) were used to select Specially Protected Areas under the Antarctic Treaty (Usher and Edwards, 1986). However, more criteria have also been used in the Antarctic including representativeness, area with unique species, areas with type localities, and areas which contain especially interesting animal populations.

In selecting areas for nature reserves, a common principle has been to establish an ecologically representative series of reserves; based on well-founded biogeographical principles to represent the major ecosystems (Mackinnon *et al.*, 1986). This principle has been applied to both National Nature Reserves and to Biosphere Reserves. In the case of Biosphere Reserves, the International Co-ordinating Council for the Man and the Biosphere Programme has identified four mandatory criteria (although they are not defined in operational terms): representativeness, diversity of ecosystems and communities, naturalness, and effectiveness of an area as a conservation unit (MAB, 1974). In addition to the establishment of a representative series, some taxonomic groups have been the focal point. For example, criteria for identifying internationally important populations of wildfowl have been readily available and have thus helped to conserve the habitats of those wildfowl (p. 134).

The variety of criteria range from those which are easily quantified (such as number of species) to those based on subjective assessments (such as fragility). An example of a simple but more objective and quantifiable approach was Goldsmith's (1987) selection procedures for forest nature reserves in Canada. The preferred criteria were: number of plant species, number of rarities, area, and diameter of the largest trees. In that method the four criteria are easily quantified and provide a good basis for ranking sites. By way of contrast, a site evaluation scheme devised by Wright (1977) has nine criteria which give ordinal values (class scores: see p. 113). The criteria are: representativeness of the ecosystem, representativeness of the geological region, habitat diversity, community diversity, species richness, species rarity, landscape category, sensitivity to disturbance, recorded history. In Wright's scheme, class scores (1–3) are awarded and then summed. The summing of class scores is not a logical procedure because the score values can represent variation which is not quantified.

No single set of criteria could be devised for all regions of the earth or for all countries. Differences in biogeography, differences in human population densities (which result in different pressures and demands on land use) and variation in scale of biological surveys are three reasons why different sets of criteria can be expected to be used in different countries. For this reason, the contrasting land use of Britain and Australia have been selected to illustrate two different approaches to site selection for nature reserves.

6.6.2 A history of site selection in Britain

An important basis for nature reserve selection in Britain was Tansley's (1939) vegetation classification and the principle of key site selection. The key site principle for nature conservation was made clear in Cmd 7122 (Ministry of Town and Country Planning, 1947) in a discussion of the main purposes of representative sites:

> to preserve and maintain as part of the nation's heritage places which can be regarded as reservoirs for the main types of community and kinds of wild plants and animals representative in this country, both common and rare, typical and unusual, as well as places which contain physical features of special or outstanding interest.

The original National Nature Reserve selection was much influenced by Tansley's classification. Later, in the 1960s, Tansley's classification was revised to include new information but the basic approach was towards plant associations. A major contribution to assessment of sites for nature conservation in the UK came in 1977 with the publication of *A Nature Conservation Review* (Ratcliffe, 1977). This two-volume publication described and assessed ten criteria for evaluation of sites as a basis for the establishment of representative examples of biological natural areas. These criteria are aimed at evaluating areas of broadly similar vegetation type. The Nature Conservation Review was also a vehicle for identification of key sites (of National Nature Reserve Standard). In all, 784 sites representing 165 000 ha (4.2% of Britain) were identified.

The criteria for key site selection have since been the subject of many critical appraisals (for example Goldsmith, 1983, 1991) and have also provided a useful basis for field studies in ecology and conservation. In outline, the ten criteria described by Ratcliffe (1977) were as follows:

1. *Area (size or extent)*. This is based on the idea that larger sites are better because they may contain more species. It is also based on the idea that there is a minimum area which needs to be safeguarded in order to maintain the conservation interest of the locality. Size (measured as number or density) has also been a main criterion for assessing highly aggregated populations of certain species.
2. *Diversity (variety or richness)*. This applies to both species and communities. However, although the general aim is to conserve areas diverse or rich in species and communities which are varied in struc-

ture, there are some interesting communities such as moorlands and heathlands which intrinsically are plant species poor and which have a uniform topography and structure. Species richness is a useful criterion but it is the number and abundance of indicator species characteristic of the particular ecosystem type which is especially relevant.

3. *Naturalness.* This is a relative and rare condition in industrialized countries but the argument is that the types of habitats least modified by humans should be given priority. Some forms of traditional management are essential to maintain some habitats: for example grazing of chalk grassland is essential for maintaining chalk grassland plant species. Evaluation and quantification of naturalness is described on p. 50).

4. *Rarity.* This criterion can be quantified in an ecological sense (p. 38) but rarity can also be thought of as an expression of value. One of the most important functions of a nature reserve is to protect rare species and rare communities. However, it is important to determine the factors which have made the species rare so that appropriate management of the species can be undertaken on the nature reserve.

5. *Fragility.* This is an indication of the degree of sensitivity of habitats, communities and species to environmental change and impacts caused by the activities of humans. In general the most fragile ecosystems and species have higher conservation priorities because they are most likely to be lost or damaged. However, it is often the case that conservation of fragile ecosystems may be difficult and often requires large resources.

6. *Typicalness.* This is based on the idea that it is necessary to represent the typical and commonplace within a field of ecological variation. Sometimes sites have to be selected for their characteristic and common habitats, species and communities. The rationale here is that the best examples of various communities or ecosystems should be selected but at the same time it is recognized that their quality may be determined by features which are unusual. Unusual communities or ecosystems may have only a few available examples, whereas there may be a much wider choice of those which are typical or common. An example of how typicalness can be quantified is given on p. 50.

7. *Recorded history.* The extent to which a site has been used for education, natural history studies and scientific research is a factor of considerable importance. Long-standing natural history and scientific records enhance understanding of conservation and may provide a basis for the analysis of ecological processes.

8. *Position in an ecological/geographical unit.* The aim here is that where practicable, and without lowering the standards of selection, it is desirable to include within a single geographical location as many as possible of the important and characteristic formations, communities and species of an area. For example, a woodland area adjoining riparian habitats may enhance the value of the site.

9. *Potential value.* This refers to sites which with appropriate and successful management or possibly even natural change show a potential for
developing features of particular value in conservation. This criterion
is an amalgamation of other criteria and suggests that habitat creation
is an important component of conservation (see p. 76).

10. *Intrinsic appeal.* This has possibly caused more controversy (from both
practical and theoretical points of view) than any other criterion. It is
an interesting philosophical criterion and whereas science may view
all organisms as equal, pragmatism dictates that in nature conservation
it is inevitable that more weight will be given to some taxonomic
groups and certain features than others. Generally speaking, birds,
colourful wild flowers and rivers may arouse more enthusiasm than
beetles, snakes or limestone pavements, but it is very necessary that
the conservation of less popular groups and features is adequate.

Outside Britain, these and two other criteria (presence of buffer zones,
representativeness) are used for selection of protected areas. There is a
possibility of some confusion between typicalness as used in the above
sense and representativeness. The latter has been used for nature reserves
which represent the range of biological variation in a given region. Representativeness can be quantified (p. 50).

It would be naive to think that site selection required only a checklist of
criteria and the inspection of sites (of broadly similar vegetation). For
reasons of space the description of the ten criteria is simplified here. In
practice the criteria would be qualified according to what biological communities were being assessed whether they be woodland, lowland grassland
(acidic, calcareous), coastland or heathland. To appreciate the applications
of the Nature Conservation Review, the detailed qualifying statement must
be consulted in the Review. Furthermore, the selection of a series of sites
is very much dependent on detailed survey work and recording and therefore it is a gradual and cumulative process.

There are 784 key sites in the Nature Conservation Review but only 234
National Nature Reserves in Britain. However, it is the large number of
Sites of Special Scientific Interest (SSSI) which helps to supplement the
number of National Nature Reserves and helps form a national network
of representative and other areas (the SSSI selection procedure can also be
directed at specific species such as dragonflies as well as just sites). Whereas
National Nature Reserves are generally owned or leased by the conservation authority, an SSSI can be owned by anyone but use and management
of the site is restricted to certain conditions. A Site of Special Scientific
interest was defined under the Wildlife and Countryside Act 1981 as an
area 'of special interest by reasons of any of its flora, fauna, or geological
or physiographic features'. It has been the NCC's statutory duty under
section 28 of the Act to notify any area of land which in its opinion is of

special interest (note the changes in government organizations in Box 6.1; such changes are disruptive and do little to foster efficiency). A major publication aimed at helping NCC staff in the selections of biological SSSIs has been *Guidelines for Selection of Biological SSSIs* (NCC, 1989). The guidelines for geological and physiographical SSSIs is to be provided separately.

The main need is to ensure that the series of SSSIs throughout the country is sufficient in kind, number and extent to conserve the total national 'special interest' of the range of variation in habitats, with their associated flora and fauna. For practical purposes it has been felt necessary to relate the selection process to geographical subdivisions known as Areas of Search (AOSs). For many habitats and species, the minimum of one example or population per AOS will not be enough and the guiding principle is that, as rarity or other special value increases, so does the need to select a larger proportion of the total remaining area or population.

The Guidelines include a paragraph devoted to caution with respect to the use of scoring systems in nature conservation evaluation. Although it is suggested that scoring systems are useful as standardized methods for ranking sites of similar types, it is also noted that scoring systems are invariably too loaded with hidden value judgements and subjective assumptions.

The criteria in the Guidelines (essentially the same as in the Nature Conservation Review) must not be seen as a shopping list set by rigid rules. In the final analysis each case rests on opinion. Also, the application of the criteria is complex and different criteria or different combinations are needed for evaluation and selection. The primary criteria are size, diversity, naturalness, rarity, fragility and typicalness whereas the secondary criteria are: recorded history, position in an ecological/geographical unit, potential value and intrinsic appeal. Some of the qualifications in the Guidelines for four of the primary criteria are as follows:

1. *Size (extent)*. Provided the intrinsic vegetational quality of the habitat is acceptable, the area must be big enough to be viable in respect of the resistance of the flora and fauna to edge effects, loss of species and colonization by invasive, unwanted species. There can be no standard rule for reasons of different requirements by different communities. The example given in the Guidelines is that whereas a fragment of fen of 100 ha may continue to dry out if surrounded by farmland, many drier meadows of just 1 ha may retain their floristic composition and interest.

2. *Diversity*. At the plant community or animal assemblage level, some are intrinsically more species rich than others and so comparisons can be made only between similar communities and assemblages. In Britain, one standard of floristic diversity has been quantified in the National

Vegetation Classification (p. 54), at least with regard to national variability (Rodwell, 1991).

3. *Naturalness*. Although habitats not modified by man are highly valued, they are now very rare in Britain. Semi-natural habitats are usually being considered but nevertheless those with gross or recent human modification would probably be of low quality. Physical modification varies and some, if part of the management procedure, may be desirable.

4. *Rarity*. This can have an important effect and can make selection against uniform minimum standards inappropriate. The more rare the habitat, the stronger the case that the qualifying standards should be adjusted downwards. For example, unimproved meadow grasslands are now so fragmented and localized that the minimal acceptable area has to be quite small if some are to be represented at all. The rarity of a species is regarded as a measure of its proneness to extinction (note the PVA underlying *ex situ* conservation on p. 12), and this concept is expressed in the variety of terms such as endangered, vulnerable and threatened.

There are, in addition to the basic primary and secondary criteria, some other important subjects in the Guidelines and including international obligations, ecological implications of fragmentation and thus the need for buffer zones, and the potential importance of ecological links between protected areas to form networks or mosaics. Under the subject of international obligations, there are notes on four conventions (all of which are mentioned elsewhere in this volume), namely the Ramsar Convention, the Bonn Convention, the Berne Convention and the World Heritage Convention.

6.6.3 Research on site selection for nature reserves in Australia

Although Australia is a large country with a small localized, resident population, the degradation, loss and fragmentation of natural habitats has followed a pattern similar to that seen throughout Europe and North America (Saunders *et al.*, 1987). Today, habitat fragmentation is a major problem which is largely restricted to the coastal and subcoastal temperate and tropical areas (i.e. where the population is concentrated and where intensive agricultural development has occurred). In other areas, especially in the arid 70% of the continent, habitat degradation due to stock grazing and feral animals, rather than fragmentation is a major cause for concern.

Concern about habitat degradation and fragmentation in Australia had its beginnings towards the end of the 19th century when professional societies first raised concern about implications of forest removal (Frawley, 1988). But it was not until the 1960s when the conservation movement began to make itself felt with a period of vigorous campaigns and the appearance of influential publications such as Marshall's *The Great Exter-*

mination published in 1966. In the two decades that followed there was a rapid growth in the number of conservation organizations and Australian environmentalism became a powerful force. However, the lead, in terms of rationalizing the needs and procedures for conservation in Australia has been undertaken by scientists in the universities and in establishments such as the CSIRO.

National park designations commenced in Australia in 1879 when the (Royal) National Park was established near Sydney. From the 1960s onwards there was a spectacular increase in the number of protected areas and by 1986, about 4.5% of land had been set aside as national parks and other conservation areas (Mobbs, 1989). As is common elsewhere, designation of nature reserves in Australia was for many years largely influenced by pressures of other land-uses and by land ownership: many were established without knowing whether they were the best ones for representing particular features (Pressey, 1990; Pressey *et al.*, 1990). However, in the early 1970s, the representativeness of nature reserves in Australia and New Guinea was addressed by Specht *et al.* (1974) and later the rationale underlying site selection for nature reserves in Australia began to be questioned and researched. In 1974, the need for a comprehensive reserve system in Australia was affirmed and the guiding principles on which it should be based were determined at a conference on 'A National System of Ecological Reserves in Australia' (Fenner, 1975). Understandably it was recognized that prerequisites for achieving a representative series of protected areas are good survey methods and data sets. However, extensive surveys and the acquisition of good field data are not easy tasks in many parts of Australia because of the area (see also p. 182 in connection with surveys for wilderness areas).

In Australia, there has been wide recognition that not all areas can be set aside as nature reserves. That recognition seems to have led to considerable efforts to ensure that the number and location of nature reserves produces the best conservation efforts. Perhaps that is the reason why, over the last 20 years, there has been much objective, quantifiable and repeatable research on site selection for reserves in Australia. That is, there has been a departure from the subjective and *ad hoc* approach (see for example Margules, 1989). However, the theory has not always been put into practice and some feel that there is a gap between the theory and practice (R. Pressey personal communication).

Research on site selection in Australia has addressed two interdependent problems: one has been representativeness and the other site selection. Research on the former has been extended to the concept of networks, that is, representative collections of remnant patches of habitats in particular regions (Margules and Nicholls, 1987). Habitat degradation in the Mulga Biogeographic Region of Southwestern Queensland (one of 13 regions in Queensland) has provided an opportunity for the application of the representativeness concept. Native grasslands occupy only a small proportion of the Mulga Region and degradation has occurred in the Mulga and Mulga-Poplar box woodlands where grassy understoreys have been replaced by native shrub species (woody weeds) as a result of overgrazing and then soil erosion (Figure 6.2). It was proposed by Purdie *et al.* (1986) that there should be a reserve network containing representatives of all plant species, vegetation types and landforms typical of the area. At the same time, thought was given to preserving regional genetic diversity and a full range of habitats with which the species are associated. The method adopted by Purdie *et al.* (1986) made use of computers for analysing existing resource maps. The basic units of analysis were 9 × 9 km² grid squares. The areas of high conservation value in terms of ecosystem diversity, representativeness, rarity and naturalness were then identified. Key areas which had the highest vegetation diversity, ecosystems in the best condition and greatest number of special features was obtained using field data from individual blocks (clusters of squares representing highest theoretical conservation value). Three major National Parks have now been designated in the Mulga region and that designation owes much to the work by Purdie and her colleagues.

The most common method of site selection for reserves has been based on a scoring system where sets of criteria with different weights are employed (see Bedward *et al.*, 1991 for a review). A criticism of this approach has been that it could lead to an imbalance in representation because areas of differing priority could contain similar sets of species, communities or habitats (Kirkpatrick, 1983). In other words there could be duplication, something which is to be avoided when resources for conservation are limited.

An alternative approach which has received considerable research interest in Australia has been the iterative (meaning stated or checked repeatedly) approach (Kirkpatrick, 1983; Pressey and Nicholls, 1989a). In one iterative approach (Kirkpatrick, 1983) in Tasmania the aim was to assess priorities for the conservation of threatened species. Basically, it is assumed that not all theoretically valuable conservation areas can be reserves and therefore there is a need to select areas. A priority list could be established but the problem still remains as to which of the reserves (the first 3%, 6% or 80%?) would result in the best conservation. If the first area (in terms of priorities) is established as a nature reserve then the status of a set of species will be improved. Designation of the next best area, rather than an area lower in the priority list, would not necessarily be most effective in terms of conservation effectiveness. That is, with establishment of the first reserve, the status of some species is assumed to be improved and this iterative approach means that there is maximum conservation per unit area set aside for nature reserves. The key points are: by virtue of the natural features they contain, sites can complement each other in a reserve network;

(a)

(b)

Figure 6.2 (a) Mulga shrubby low open woodland (*Acacia aueura* with shrub stratum of *Eremophila bowmannii*, *Dodonaea petiolaris* and *Cassia sturtii*). East of Quilpie, Queensland, Australia. (b) *Acacia aueura* grass low open woodland, in 'Aldville', Queensland, Australia. Photographs kindly provided by Rosemary Purdie.

and iterative approaches readjust the potential contributions of unreserved sites to a network each time a site is added to the network.

The earlier research of Kirkpatrick (1983) has now been taken further by Margules *et al.* (1988) and Pressey and Bedward (in press). Margules and Pressey are two scientists who currently lead the discussion about reserve selection theory in Australia by way of their attempts to maximize the conservation of biodiversity with selection of networks of reserves.

But is it true that the iterative method for establishing priorities for the selection of nature reserves is better than scoring? The scoring and iterative approaches to conservation evaluation have been researched by Pressey and Nicholls (1989b) on the basis of a measure of 'efficiency' (defined as a measure of the effectiveness of the procedures for conservation evaluation in promoting the basic conservation goal). On the assumption that there are limitations as to the number and location of nature reserves that can be established in any region, Pressey and Nicholls (1989b) considered how various scoring criteria influence the number and area of highest ranking sites needed to contain all attributes a given number of times. Efficiency of sampling was defined by Pressey and Nicholls (1989b) as follows:

$$E = 1 - (X/T)$$

where E is efficiency (varies from 0 to 1), X is the number or extent of highest ranking sites needed to contain all attributes a given number of times, and T is the total number of area of sites. Efficiency values were calculated for two data sets by way of scoring criteria and by iterative analysis. The individual criteria for scoring were based on rarity, representativeness, diversity and area. In general, they found that scoring criteria varied in their efficiencies and were consistently less efficient than the iterative analyses.

There has been some concern about the fact that Margules *et al.* (1988) do not take into account land condition and the regional concept. For example R. Purdie (personal communication) has noted that an area of land shown theoretically to be of high conservation priority may be low in priority for land acquisition if it has been severely overgrazed and degraded. Also, attempting to include every species in the 'study region' may not be a realistic aim or efficient. In the case of the study by Purdie *et al.* (1986), it was felt that land systems/vegetation types which were not or only poorly represented in the proposed reserve were possibly better conserved in adjacent biogeographical regions. Therefore, the efficiency calculations of Pressey and Nichols regarding scored criteria versus iteration, although theoretically quite correct, may not account for some of the practical problems.

6.7 THE BENEFITS AND LIMITATIONS OF PROTECTED AREAS

Protected areas have many benefits. Some protected areas which span political boundaries help towards a stable coexistence and some act as security zones supported by military planners (MacKinnon *et al.*, 1986). National parks provide facilities for tourism, recreation and education. Some nature reserves are reservoirs of wild species which may provide genetic material for improvement of crop and grazing plants. Beneficial insects (such as predators of insect pests) and pollinators of fruit trees and other crop plants may depend on nature reserves for their existence. Ecological studies, educational studies and environmental research depends on and also benefits from the existence of protected areas. Studies of relatively undisturbed ecosystems are helped by the existence of nature reserves. Very little is known about ecosystem processes and the ecology of change on a large scale. Nature reserves provide habitats where ecological change can be monitored and could also provide a network of sites for monitoring effects of pollution (from effects of acid rain to the fate of radionuclides). Some larger national parks, wilderness areas and biosphere reserves may soon be the only places on earth suitable for long-term studies of ecological and landscape change. Perhaps of greatest overall importance, the designation of protected areas helps to slow the rate at which natural areas are damaged, reduced, fragmented and isolated.

Although about 4% of the world's land surface has been set aside for protection of plants and animals, the designated nature reserves are not representative of the world's ecosystems, most of the nature reserves are isolated and are not located where there is a greatest need to conserve. Most nature reserves are designated in areas which serve no other purpose. Some have been established so as to save the last remaining fragments of a devastated landscape. For example, it has been suggested by many experts (WCED, 1987) that 20% of tropical forest should be protected but to date only 5% has been given any protection and some of that protection is inadequate because exploitation of the forest continues in various forms.

If there are not enough protected areas, what can be done to persuade decision makers to designate more? There has to be an incentive. There are few countries in the world where protected areas can be established without incentives. After all a fundamental obstacle to establishment and tenure of nature reserves is loss of revenue and the perceived loss of potential commercial developments. In Britain, incentives of an interesting kind have been devised to encourage landowners to conserve certain land-scapes (Peak National Park, 1990). Project maintenance grants of up to £3000 per ha are made for the management and retention of flower-rich fields and riverside meadows. English Nature have also introduced a programme whereby payments are to be made to landowners for improving the conservation value of SSSI (The Wildlife Enhancement Programme).

Perhaps the potential economic value of nature reserves in general should be promoted wherever possible, using suitable incentives. Although it is not easy to reconcile conservation with economic incentives, there have been a number of attempts (McNeely, 1988). Direct incentives include those in cash (subsidies for reforestation, research grants, World Heritage Fund grants), and those in kind (food for work in a reserve, forest concessions and conservation-sponsored equipment). Indirect incentives include fiscal measures (compensation for damage by wild animals, commodities agreements and debt swaps), provision of services and social factors (enhanced land tenure, training for staff and international databases).

Even if there are incentives and designations taking place the protected area has to be managed. In other words the principle of designation may be accepted but the investment in terms of managers and management plans can be lacking. In too many parts of the world, protected areas for conservation of nature and landscape exist in name only. In many parts of the world there is a lack of sufficiently strong legislation and law enforcement to maintain the integrity of the reserve. The problem of enforcing protection has been particularly difficult in South Africa where although many reserves have been designated, there has been little in the way of support for the surveillance of those reserves let alone the ecological research necessary for their management.

The relatively small areas of many national nature reserves and national parks has important implications for the plants and animals in the protected area. In general, the areas are so small that management has to take place instead of leaving nature to take its course. In other words, without management, there may be large fluctuations in the population size of some species; populations of grazing animals exceeding the carrying capacity of the reserve is a common problem in several countries. Inbreeding and loss of genetic variation in small isolated populations in nature reserves is becoming an increasingly common problem. Furthermore, many nature reserves are too small to absorb the effects of disturbance and damage caused by invasive species, pollution and erosion. One obvious but perhaps unrealistic solution is to extend the area of existing nature reserves. Alternative solutions are the management of biodiversity in the areas between reserves, and active management of public lands to protect their biodiversity. This latter suggestion would require a whole new approach to land planning and certainly land evaluation.

An interesting assessment of the area of reserves in western North America (Newmark, 1985) was based on the legal and biotic boundaries (the area encompassing the entire watershed of a park, enlarged if necessary to maintain a 'minimum viable population' for the terrestrial non-flying species with the largest home range found within the legal boundaries). Eight national parks were examined and seven of the eight were found to be too small by factors ranging from 6 to 100. Although none of the parks was originally designated for conservation of mammals, this difference between legal and ecological boundaries does raise an important problem.

Many nature reserves are remnants of larger natural areas. That may result in the isolation of plant and animal populations and indeed some protected areas may be too small to support all the resources required by them. Nature reserves can be surrounded by many kinds of land uses and this can result in damage by pollution and physical disturbance of communities on the reserve. Invasive species from the surrounding area may colonize nature reserves, for example large expanses of rhododendrons and bracken (*Pteridium aquilinum*) have created formidable problems on many nature reserves and national parks in England and Ireland. The principle of key site selection has provided a sound foundation for the establishment of national nature reserves but that principle should be reviewed from time to time. In New Zealand, for example, the effectiveness of the reserve systems has been the subject of very useful appraisals resulting in new initiatives for the future (Norton, 1989).

There have been many arguments about the benefits of either large protected areas or several small protected areas (Single Large or Several Small – SLOSS debate). There can be no general rule because different nature reserves are established for different reasons. Support for large reserves has often not been derived from studies of the need to conserve biodiversity but has resulted from mistaken connections between the ecology of islands and the conservation of wildlife on insular habitats (Spellerberg, 1991). Nature reserves are often surrounded by varied landscapes which prevent some plants and animals dispersing. However, problems of isolation may be overcome by the establishment of linear features ('wildlife corridors') between reserves. In Western Australia, the use of corridors has been extended to a network of reserves interconnected by various linear features such as lines of trees, scrub covered banks or river edges.

Protected area buffer zones are areas which help to diminish restrictions of area and may help to reduce impacts on the protected area. There are basically three types: (1) an area which extends the habitats for some species thus allowing larger breeding populations (Mackinnon *et al.*, 1986); (2) ecological buffer zones which help to preserve the ecological integrity of the protected area; (3) socio-buffers which allow combined use by local people and conservation of wildlife. There is much research yet to be done on buffer zones but one particularly interesting application of their use has been the zone established around Cameroon's Korup National Rainforest Park which buffers the Park against unfavourable disturbance and at the same time provides resources for 30 000 local farmers.

6.8 CONCLUSIONS

The evaluation of landscapes has attracted much research on methods and on variables which affect judgements. For example, support for the use of surrogate material in order to avoid costly field surveys has resulted in work on the use of drawings, photographs and computer images. In one investigation it was found that there was a decreasing level of agreement with increasing level of abstraction with line drawings as surrogates (Tips and Savasdisara, 1986b). Measurement of attitudes and feelings about landscapes has prompted very wide discussions as has the question of how best to measure attributes of the landscape. Perhaps more research on application rather than theory is now called for and one area worthy of attention would be landscape classification. However, it is possibly more important to share experiences of landscape management and aim for successful management programmes. With so much experience around the world, it was very timely that The International Centre for Protected Landscapes was established in 1990. The task of that organization is to collect international experience of landscape protection and then to coordinate new conservation and development initiatives. But management seems to be the key and indeed the practice of improved management seems to have been adopted in several countries. For example, in Australia, the National Parks and Wildlife Conservation Act 1975 highlighted management as being important (Australian Conservation Foundation, 1975). In North America, research on the management of wilderness areas has led to the concept of Limits of Acceptable Change (p. 182).

Of the many and varied criteria used for selection of nature reserves, some have been widely used and well received. For example the criteria in the Nature Conservation Review (Ratcliffe, 1977) have become widely accepted in Britain, despite the sometimes critical discussion surrounding those criteria (see examples in review by Goldsmith, 1983). A total of 15 criteria based on Ratclife's criteria have been put forward for consideration in the IUCN publication *Managing Protected Areas in the Tropics* (Mackinnon *et al.*, 1986). Some of the additional criteria were genetic conservation, indispensibility (areas which protect resources such as water catchments), and modified landscapes augmenting biological values (national or cultural sites).

In the application of these criteria, it would be expected that there would be some differences in agreement on site values and also some difference of opinion about the relative importance of the different criteria. In one analysis of nine people's evaluation of potential conservation sites, Margules and Usher (1984) found that there were indeed difficulties in arriving at fixed weights for criteria of evaluation, thus justifying the fairly flexible approach advocated by Ratcliffe (1977). Margules and Usher also found that there was considerable agreement on the relative importance of each criterion. This is important and it enabled the authors to suggest a possible hierarchical classification of the criteria used for evaluation of sites for nature reserves. For small sites, stage one included ecological fragility followed by threat of human interference, species rarity, habitat rarity or uniqueness. Stage two was diversity and stage three was management factors and scientific value. For large sites, stage one included representativeness followed by area, position in the geographical/ecological unit, naturalness and wildlife reservoir potential. The second stage was diversity and the third stage was management factors and scientific values. Having considered these analyses of Ratcliffe's 10 criteria, it would seem unjust not to conclude that in general terms these criteria provide the most useful basis for evaluation of specific areas for designation as nature reserves – but only in countries which have surveyed the areas and have good data sets.

In countries which do not have good data sets from natural areas, more general methods for allocation of land for protection are called for. For example, a basis for allocation of land in general terms can be embodied in broadly worded legislation. Similarly, the basis for selection of large protected areas such as biogenetic reserves may be found in the objectives of establishing those reserves. Another approach is by an agreed management programme. That is, rather than first identifying the precise area by way of criteria or objectives in legislation, an alternative approach is to allocate large areas of land via landowners and, provided they are willing partners, there can be an agreement to manage the land in a certain manner. This approach has been usefully described by Horwich's (1990) work on black howler monkeys (*Alouatta pigra*) in Belize. This work was directed at the basis for the establishment of a protected area for that species which was to contact the local people, initiate an educational programme and collect information on landownership. It was then possible to approach and consult with each landowner in turn with the objective of obtaining an agreement from the landowner about some form of management. This approach avoids the danger of ending up with widely scattered protected areas. However, it is dependent on collaborative efforts of the landowners and there may be long-term difficulties if landownership changes.

Protected areas are essential but emphasizing the methods for selection of protected areas tends to focus attention away from the limitations of existing protected areas. There has been much debate about how much land should be set aside as protected areas; 5, 8, 10 or more per cent. It does not help to try and devise one overall, global figure because of differences in population density, differences in biogeography and differences in ecosystems. We should be trying to set aside as much as possible within the broad range of categories and types of protected areas. In other words the range of protective areas with the different objectives should be used to maximize as much protection as possible.

Limitations of existing protected areas arise from inadequate representation of natural ecosystems, insufficient protection and ecological implications of fragmentation and isolation. It is naive to believe that any existing series of nature reserves (whether they be based on climax communities or seminatural and plagioclimax communities) are representative. Indeed the protected areas of the world are not representative of all the different ecosystems. The security of protected areas is often in question, some being deliberately damaged and some being lost to other land uses when a lease has expired. Many protected areas are fragments of larger ecosystems, too small to support some species and are surrounded by different land uses; consequently the wildlife on reserves can be affected by the surrounding land and activities on that land (see discussion about effects of fragmentation on p. 12). It would seem important therefore that we try to overcome some of these limitations.

For example, there does not appear to be enough large protected areas. In one examination of protected area size in north America, along with an appraisal of viable population theories, Grumbine (1990) concluded that biodiversity in national parks and forests was inadequately protected. Boundaries as well as area of protected areas have been subject to criticism. That is, if the surrounds of a protected area have important implications for conserving both landscapes and biodiversity, more attention needs to be given to the designation of boundaries to ensure that they take into account existing or potential land-uses as well as features likely to ensure the continued integrity of the protected area (Lucas, 1992). Boundaries need not be lines but can take the form of areas of land in the form of buffer zones (buffering the effects of invasive species, physical damage and pollution). The nature, design and management of buffer zones may have important implications for protected areas and they have possibly more value than was previously thought. However, buffer zones and linear features (some of which may be wildlife corridors) must not be used as an argument for continued loss and fragmentation of natural communities nor should buffer zones and corridors be used as an argument to support the concept of nature reserves. With increasing rates of insularization the world over, it is perhaps time to look at protected areas in a much broader context and on a much larger scale.

The plants and animals in some protected areas are isolated and subject to disturbance from the surrounding land. That being the case, planning future strategies for the land should pay more attention to the needs of conserving biodiversity and not put conservation low in the priorities of land use. Conservation by way of protected areas needs to be put in a wider context of varied land uses: for example when designating a protected area, surrounding buffer zones and corridors linking similar areas could be designated and managed appropriately.

Obviously the designation of protected areas is not the end of the process.

There has to be management of those reserves but of equal importance is the need to undertake a continuous process of monitoring the status of the protected areas and also of evaluating the effects of management. The need for monitoring the status of protected areas could well apply to the SSSI network in Britain which has been less than successful because of loopholes in the law and because of the growing number of damaged sites (RSNC, 1989; Adams, 1991; RSPB, 1991). Evaluating management effectiveness is very important but slightly beyond the scope of this book. Checklists have been devised for evaluating management and examples in the IUCN publication *Managing Protected Areas in the Tropics* are well worth considering.

REFERENCES

Adams, B. (1991) SSSI's: who cares? *ECOS*, **12**, 59–64.

Australian Conservation Foundation (1975) *Landscape Conservation*. Australian Conservation Foundation, Melbourne.

Bedward, M., Pressey, R.L. and Nicholls, A.O. (1991) Scores and score classes for evaluating criteria: a comparison based on the cost of reserving all natural features. *Biological Conservation*, **56**, 281–94.

Curry-Lindahl, K. (1988) The global role of national parks for the world of tomorrow, in *Conservation of Hope*, The Horace M. Albright conservation lectures, The Regents of the University of California, Berkeley, University of Idaho Press, Moscow, Idaho, pp. 263–316.

Dearden, P. and Sadler, B. (1989) *Landscape Evaluation: Approaches and Applications*. University of Victoria, Victoria, Canada.

Fenner, F. (ed.) (1975) *A National System of Ecological Reserves in Australia*. Reports of the Australian Academy of Science, No. 19, Canberra.

Fines, K.D. (1968) Landscape evaluation: a research project in East Sussex. *Regional Studies*, **2**, 41–5.

Frawley, K. (1988) The history of conservation and the national park concept in Australia: a state of knowledge review, in *Australia's Ever Changing Forests* (eds K.J. Frawley and N.M. Semple) ADFA, Geography & Oceanography Special Publication no. 1, pp. 395–418.

Gilg, A.W. (1974) A critique of Linton's method of assessing scenery as a natural resource. *Scottish Geographical Magazine*, **90**, 125–9.

Gilg, A.W. (1976) Assessing scenery as a natural resource: causes of variation in Linton's method. *Scottish Geographical Magazine*, **92**, 41–50.

Goldsmith, F.B. (1983) Evaluating nature, in *Conservation in Perspective* (eds A. Warren and F.B. Goldsmith) John Wiley, Chichester, pp. 233–46.

Goldsmith, F.B. (1987) Selection procedures for forest nature reserves in Nova Scotia, Canada. *Biological Conservation*, **41**, 185–201.

Goldsmith, F.B. (1991) The selection of protected areas, in *The Scientific Management of Temperate Communities for Conservation*, (eds I.F. Spellerberg, F.B. Goldsmith and M.G. Morris), Blackwell, Oxford.

Götmark, F. and Nilsson, C. (1992) Criteria used for the protection of natural areas in Sweden 1906–1986. *Conservation Biology*, 6, 220–231.

Grumbine, E. (1990) Viable populations, reserve size, and Federal lands management: a critique. *Conservation Biology*, 4, 127–34.

Horwich, R.H. (1990) How to develop a community sanctuary – an experimental approach to the conservation of private lands. *Oryx*, 24, 95–102.

Kirkpatrick, J.B. (1983) An iterative method for establishing priorities for the selection of nature reserves: an example from Tasmania. *Biological Conservation*, 25, 127–34.

Leopold, A. (1921) The wilderness and its place in forest recreational policy. *Journal of Forestry*, 19, 718–21.

Leopold, L.B. (1969) Quantitative comparison of some aesthetic factors among rivers. *US Geological Survey Circular* 620. Washington, DC.

Lesslie, R. (1991) Wilderness survey and evaluation in Australia. *Australian Geographer*, 22, 35–43.

Lesslie, R.G., Mackey, B.G. and Preece, K.M. (1988a) A computer-based method of wilderness evaluation. *Environmental Conservation*, 15, 225–32.

Lesslie, R.G., Mackey, B.G. and Schulmeister, J. (1988b) National Wilderness Inventory: stage II – Wilderness quality in Tasmania. A report to the Australian Heritage Commission 1988.

Lesslie, R.G. and Taylor, S.G. (1985) The wilderness continuum concept and its implications for wilderness preservation policy. *Biological Conservation*, 32, 309–33.

Linton, D.L. (1968) The assessment of scenery as a natural resource. *Scottish Geographical Magazine*, 84, 219–38.

Lucas, P.H.C. (1992) *Protected Landscapes: a Guide for Policy Makers and Planners*. Chapman & Hall, London.

MAB (1974) Criteria and guidelines for the choice and establishment of biosphere reserves. *MAB report series* No. 22, MAB, Paris.

MacKinnon, J., MacKinnon, K., Child, G. and Thorswell, J. (1986) *Managing Protected Areas in the Tropics*. IUCN, Gland, Switzerland.

Margules, C.R. (1989) Introduction to some Australian developments in conservation evaluation. *Biological Conservation*, 50, 1–11.

Margules, C.R. and Nicholls, A.O. (1987) Assessing the conservation value of remnant habitat 'islands': Mallee patches on the Western Eyre Peninsula, South Australia, in *Nature Conservation: the Role of Remnants of Native Vegetation* (eds D.A. Saunders, G.W. Arnold, A.A. Burbidge and J.M. Hopkins) Surrey Beatty, Melbourne.

Margules, C.R., Nichols, A.O. and Pressey, R.L. (1988) Selecting networks of reserves to maximise biological diversity. *Biological Conservation*, 43, 63–76.

Margules, C.R. and Usher, M.B. (1984) Conservation evaluation in practice. I. Sites of different habitats in north-east Yorkshire, Great Britain. *Journal of Environmental Management*, 18, 153–68.

Marshall, A.J. (ed.) (1966) *The Great Extermination*. Heinemann, Melbourne.

McCloskey, J.M. and Spalding, H. (1989) A reconnaissance-level inventory of the amount of wilderness remaining in the world. *Ambio*, 18, 221–7.

McNeely, J.A. (1988) Economics and biological diversity: Developing and using economic incentives to conserve biological resources. IUCN, Gland, Switzerland.

Mobbs, C.J. (ed.) (1989) Nature conservation reserves in Australia (1988) Occasional paper No. 19, Canberra: Australian National Parks and Wildlife Service.

NCC (1989) *Guidelines for Selection of Biological SSSIs.* NCC, Peterborough.

Newmark, W.D. (1985) Legal and biotic boundaries of western North American national parks: a problem of congruence. *Biological Conservation*, 33, 197–208.

Norton, D.A. (1989) *Management of New Zealand's Natural Estate.* New Zealand Ecological Society, Christchurch.

Peak National Park (1990) Two villages two valleys. The Peak District integrated rural development project 1981–88. Peak Park Joint Planning Board.

Penning-Rowsell, E.C. (1981) Fluctuating fortunes in gauging landscape value. *Progress in Human Geography*, 4, 25–41.

Pressey, R.L. (1990) Reserve selection in New South Wales: where to from here? *Australian Zoologist*, 26, 70–5.

Pressey, R.L. and Bedward, M. (19??) Land classifications are necessary for conservation planning but what do they tell us about the fauna? *Australian Zoologist*, in press.

Pressey, R.L., Bedward, M. and Nicholls, A.O. (1990) Reserve selection in mallee lands, in *The Malle Lands: A Conservation Perspective* (eds J. C. Noble, P.J. Joss and G.K. Jones) CSIRO, Australia, pp. 167–78.

Pressey, R.L. and Nicholls, A.O. (1989a) Application of a numerical algorithm to the selection of reserves in semi-arid New South Wales. *Biological Conservation*, 50, 263–78.

Pressey, R.L. and Nicholls, A.O. (1989b) Efficiency in conservation evaluation: scoring versus iterative approaches. *Biological Conservation*, 50, 199–218.

Purdie, R.W., Blick, R. and Bolton, M.P. (1986) Selection of a conservation reserve network in the Mulga biogeographical region of South-Western Queensland. *Biological Conservation*, 38, 369–84.

Ratcliffe, D.A. (1977) *A Nature Conservation Review.* 2 Vols. Cambridge University Press, Cambridge.

Reid, W.V. and Miller, K.R. (1989) *Keeping Options Alive: the Scientific basis for conserving biodiversity.* World Resources Institute, Washington.

Ribe, R.G. (1986) A test of uniqueness and diversity visual assessment factors using judgement-independent measures. *Landscape Research*, 11, 13–18.

Rodwell, J.S. (ed.) (1991) *British Plant Communities.* Vol. 1. *Woodlands and Scrub.* Cambridge University Press, Cambridge.

RSNC (1989) *Losing Ground.* The Royal Society for Nature Conservation, London.

RSPB (1991) *Action for Birds and the Environment. The RSPB Environmental Manifesto for the 1990s.* Royal Society for the Protection of Birds, Sandy, Bedfordshire, UK.

Stankey, G.H., Cole, D.N., Lucas, R.C., Petersen, M.E. and Frissell, S.S. (1985) The Limits of Acceptable Change (LAC) system for wilderness planning. *General Technical Report* INT–176, US Dept. of Agriculture, Forest Service, Intermountain Forest and Range Experimental Station, Ogden, UT.

Stankey, G.H., McCool, S.F. and Stokes, G.L. (1986) Limits of acceptable change: a new framework for managing the Bob Marshall Wilderness complex. *Proceedings of the National Wilderness Research Conference: Current Research.* US Department of Agriculture, General Technical Reports, pp. 526–30.

Saunders, D.A., Arnold, G.W., Burbidge, A.A. and Hopkins, A.J.M. (1987) *Nature Conservation – the Role of Remnants of Native Vegetation*. Surrey Beatty, Melbourne.

Shafer, E.L., Hamilton, J.F. and Schmidt, E.A. (1969) Natural landscape preferences: a predictive model. *Journal of Leisure Research*, **1**, 1–9.

Smith, P.G.R. and Theberge, J.B. (1986a) Evaluating biotic diversity in environmentally significant areas in the Northwest Territories of Canada. *Biological Conservation*, **36**, 1–18.

Smith, P.G.R. and Theberge, J.B. (1986b) A review of criteria for evaluating natural areas. *Environmental Management*, **10**, 715–34.

Specht, R.L., Roe, E.M. and Boughton, V.H. (1974) Conservation of major plant communities in Australia and Papua New Guinea. *Australian Journal of Botany*, Suppl. 7.

Spellerberg, I.F. (1991) *Monitoring Ecological Change*, Cambridge University Press, Cambridge.

Tansley, A.G. (1939) *The British Islands and Their Vegetation*. Cambridge University Press, Cambridge.

Tips, W.E.J. and Savasdisara, T. (1986a) The influence of the socio-economic background of subjects on their landscape preference evaluation. *Landscape and Urban Planning* 13, 225–30.

Tips, W.E.J. and Savasdisara, T. (1986b) Landscape preference evaluation and the level of abstraction presentation media. *Journal of Environmental Management*, **22**, 203–13.

Usher, M.B. (1980) An assessment of conservation values within a large Site of Special Scientific Interest in North Yorkshire. *Field Studies*, **5**, 328–48.

Usher, M.B. (1986a) Insect conservation: the relevance of population and community ecology and of biogeography. *Proceedings of the 3rd European Congress of Entomology*, pp. 387–98.

Usher, M.B. (ed.) (1986b) *Wildlife Conservation Evaluation*. Chapman and Hall, London.

Usher, M.B. and Edwards, M. (1986) The selection of conservation areas in Antarctica: an example using the arthropod fauna of Antarctic islands. *Environmental Conservation*, **13**, 115–22.

WCED (1987) *Our Common Future*. Oxford University Press, Oxford.

World Wide Fund for Nature (1991) *Conservation of Tropical Forests and Biodiversity*. WWF, Godalming, Surrey.

Wright, D.F. (1977) A site evaluation scheme for use in the assessment of potential nature reserves. *Biological Conservation*, **11**, 293–305.

—7

Evaluation and assessment in planning and development

7.1 INTRODUCTION

In most industrialized countries, planning procedures must take into consideration the effects of projects and changes in land use on human health and welfare. Some procedures take into account the effect on the natural environment. Concern for the effects of projects on the natural environment has grown over the last few years and that has brought with it changes in legislation requiring the effects on the natural environment to be avoided wherever possible, monitored and audited. Methods of environmental assessment which are appropriate to planning procedures and new projects have been developed slowly over the last 20 years and are therefore not new. This is not to say that there is no longer any new and exciting innovations. More recently, the increasingly widespread use of statutory environmental impact assessments has provided some excellent opportunities for developing new, quick and cost-effective ways of evaluating areas of conservation importance and assessing likely effects of projects on plants, animals, habitats and ecosystems.

7.2 EVALUATION AND LAND-USE PLANNING

7.2.1 Views of ecologists

Ecological evaluation in planning is based on and makes use of land-use mapping. It is useful therefore to look first at one of the earliest, most simple yet very effective methods of ecological evaluation as used in planning. In Hampshire (southern England), evaluation maps have been widely used in planning exercises. For example, Tubbs and Blackwood (1971) devised an evaluation method which was basically a map exercise (relative ecological evaluation map). The region being considered is divided into 'ecological zones' (Box 7.1) and given a score according to specific criteria from a scale ranging from 1 to 5. The 'ecological zones' are characterized by a dominance of one of three broad categories of either agricultural land,

Box 7.1 An ecological evaluation method for planning purposes shown for an area in the south of England. From Tubbs and Blackwood (1971) with kind permission of C. Tubbs.

Ecological Zone Type	*Relative value*
Unsown vegetation (including non-plantation woodland)	Category I or II, depending on (at present) subjective estimate of rarity of habitat type and presence of features of outstanding scientific importance.
Plantation woodland	Category II or III, depending on an (at present) subjective estimate of value as wildlife reservoirs.
Agricultural land	It is assumed that the relative value of agricultural zone will depend on the extent of habitat diversity and that this will be reflected in the presence and amount of a number of definable features. These have been grouped and evaluated according to a system of subjective scoring, as follows:

Scoring for each group of features

0 = None or virtually none present in zone;
1 = Present (not conspicuous feature of zone);
2 = Numerous (conspicuous feature of zone);
3 = Abundant.

Evaluation of zone

Scoring	*Category*
15–18	II
11–14	III
6–10	IV
0–5	V

Southern England

Grouping of features

1. Permanent grassland;
2. Hedgerows and hedgerow timber;
3. Boundary banks, roadside cuttings and banks, roadside verges;
4. Park timber and orchards other than those in commercial production;
5. Ponds, ditches, streams, and other watercourses;
6. Fragments of other unsown vegetation (including woodland) smaller than the one-half kilometre square size criterion.

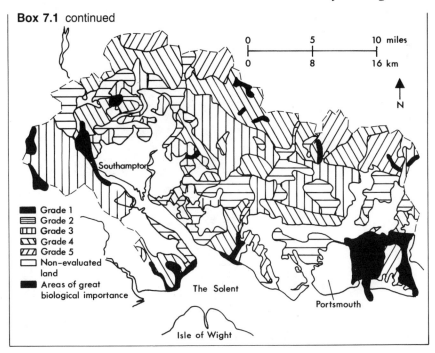

Box 7.1 continued

plantation woodland or 'unsown' vegetation. Plantation woodland and unsown vegetation of more than 0.5 km² (25 ha) and which are of general interest are marked as zones on the map. Smaller areas of agricultural land are merged with the plantation woodland or unsown vegetation zones. The score for the agricultural land is based on the diversity of habitats (Box 7.1).

This method of evaluation was used in the preparatory stages of a series of structure plans for the County of Hampshire and for the Isle of Wight in the 1970s. Somewhat less rigorously prepared evaluations have been widely used in similar contexts in many other counties of England. Twenty years on (1991) this kind of evaluation is still considered to be a most helpful and a fast way of identifying the relative value of land for conservation (C. Tubbs, personal communication). However, although suitable at a county level (i.e. the strategic planning level where planners are looking for opportunities/constraints), the method is considered to be insufficiently precise for local or district plans where there is a need for habitat maps.

The Tubbs and Blackwood method was later used by the Dorset Trust for Nature Conservation (DTNC) when they prepared a habitat evaluation map showing the grading of habitat types on the Isle of Purbeck (Figure 7.1) in the County of Dorset (Wise, 1977). Dorset, especially the eastern region, is an area of both very diverse land use and conservation interest. The lowland heathlands of that area are of ecological and conservation

Figure 7.1 Aerial photograph of part of the Isle of Purbeck, south of England. Land use here is varied: e.g. agriculture, forestry, tourism, recreation, Ministry of Defence ranges, mineral extraction (clay, stone and sand), gas and petroleum extraction as well as nature conservation. Consequently the deciduous woodlands (some are ancient woodlands) and the heathlands (high conservation value) have largely been destroyed by agriculture and forestry, leaving only a few remaining fragments. Corfe Castle is shown here and in Box 7.2. Photograph supplied by BP Exploration Ltd.

interest and have been and continue to be at the centre of many disputes. There has been a need therefore to prepare maps showing the more important and more sensitive habitats (see also p. 222). The DTNC habitat evaluation map showed graded habitat types as indicated in Box 7.2.

If nothing else, the ecological evaluation map prepared for the DTNC served to confirm the exceptional conservation value of much of the area as perceived in 1977, leading to the conclusion that the whole of the Isle of Purbeck is of such ecological richness and importance that it should be

Box 7.2 An adaptation of the Tubbs and Blackwood method of ecological evaluation for the Isle of Purbeck (see Figure 7.1). Reproduced with permission of the Dorset Trust for Nature Conservation.

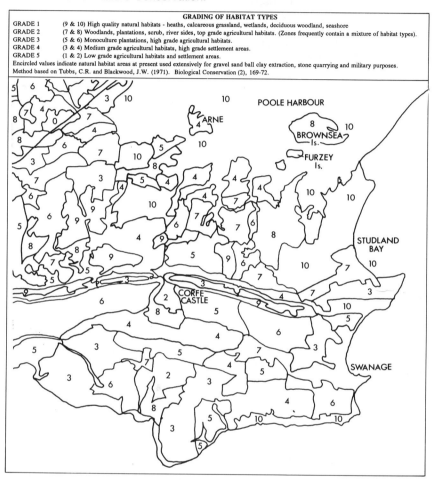

GRADING OF HABITAT TYPES

GRADE 1   (9 & 10) High quality natural habitats - heaths, calcareous grassland, wetlands, deciduous woodland, seashore
GRADE 2   (7 & 8) Woodlands, plantations, scrub, river sides, top grade agricultural habitats. (Zones frequently contain a mixture of habitat types).
GRADE 3   (5 & 6) Monoculture plantations, high grade agricultural habitats.
GRADE 4   (3 & 4) Medium grade agricultural habitats, high grade settlement areas.
GRADE 5   (1 & 2) Low grade agricultural habitats and settlement areas.
Encircled values indicate natural habitat areas at present used extensively for gravel sand ball clay extraction, stone quarrying and military purposes.
Method based on Tubbs, C.R. and Blackwood, J.W. (1971). Biological Conservation (2), 169-72.

preserved intact for all times. The same method was then extended to urban areas of south-east Dorset. The evaluation and grading of habitats is still considered to be an integral and unavoidable part of the work of the DTNC and indeed such grading has a part to play in almost every aspect of the Trust's sphere of involvement from nature reserve acquisition to development control (L. Haskins personal communication). However, the DTNC habitat evaluation map with its five categories and broad boundaries is considered today to have more detailed site boundaries and less detail in category divisions than is now required to promote the Trust's work. For this reason, the DNTC now generally uses a simple four category system where individual site boundaries are tightly defined:

Category 1. Sites of Special Scientific Interest (SSSIs)
Category 2. Sites of very high interest
Category 3. Sites of some interest
Category 4. Sites of limited or no interest

The category 1 sites can be quantified using the now well-accepted criteria for SSSIs. If the category 2 sites are to be accepted as valid categories for land use planning, they too need to be defined by recognizable criteria. The DTNC, following the example of some other wildlife trusts, are currently drawing up such criteria with which to define the category 2 sites. Non-statutory but widely recognized County 'wildlife sites' or 'Sites of Nature Conservation Interest' should then emerge.

Much of the evaluation work undertaken by some environmental consultants in Britain is similar in that they use four categories:

1. SSSIs of national or regional significance
2. Sites of regional/county importance
3. Sites of district importance (most being ancient or semi-natural vegetation)
4. Poor quality semi-natural vegetation or vegetation of some diversity but recent in origin ('good wildlife habitats')

Rather than identify areas of conservation or ecological interest, an alternative is to identify potential ecological damage caused by projects such as roads. Conflicts about the best routes for new roads (especially by-pass roads) are increasingly reported in the media and there seems to be a growing interest in how to choose between natural areas and how to quantify the costs and benefits of conserving one area while another area is destroyed. From a conservation point of view is there any generally acceptable way of assessing the least damaging route for a new road? This question was addressed nearly 20 years ago by Yapp (1973) who suggested that it would be possible to modify the method previously used by Tubbs and Blackwood (1971) for the purposes of quantifying ecological damage. It was assumed that, other things being equal, ecological damage caused by road construction would be directly proportional to its length. Therefore Yapp simply reversed Tubbs and Blackwood's scoring method so that class 5 is the best landscape where most damage would occur. For assessing agricultural land Yapp used the simple method of scoring shown in Table 7.1; examples of the calculation are shown in Table 7.2. A log scale was also suggested by Yapp but rather than using log base 10 he suggested log base 2 (i.e. 1,2,4,8,16,32) would be more realistic.

Taken in isolation as the sole ecological effect of constructing a road, Yapp's method would seem to be rather limited in its application. However, if used alongside other methods for assessing the effects of constructing a new road, then this would appear to be a useful, qualitative method of

Table 7.1 Yapp's modification of the Tubbs and Blackwood method of scoring the value of agricultural land. (From Yapp, 1973)

Score	Class	
	T & B	Yapp
15–18	II	4
11–14	III	3
6–10	IV	2
0–5	V	1

Table 7.2 Two examples of Yapp's method of determining the extent of damage along proposed roads. (From Yapp, 1973)

1 km	2 Type	3 Features	4 Class	Units of damage (Columns 1 ×4)
Route 7	Agricultural land	1+2+1+3+1+1=9	2	14
A 5	Planted woodland on one side, lake with secondary regeneration and foreshore on the other		4	20
1	Existing new road		0	0
1	Urban		0	0
2.5	Good semi-natural deciduous woodland, in part a scheduled site of special scientific interest		5	12.5
3	Existing new road		0	0
3	Part planted woodland, part bog		4	12
6	Agricultural land	2+3+2+0+0+2=9	2	12
				70.5
Route 15	Agricultural land	1+1+0+1+1+0=4	1	15
B 17	Agricultural land	3+2+3+1+1+0=10	2	34
				49

estimating ecological damage. For some taxonomic groups, roads are barriers which affect dispersal yet roadside edges can be valuable habitats and in some instances roadsides may provide corridors for dispersal for some species (Spellerberg and Gaywood, 1991). The long-term effects of roads and road traffic on ecosystems has not been well researched (apart from the effects of salt on roadside flora (Rutter and Thompson, 1986)); for example little is known about the input of pollutants from roads into ecosystems and how those pollutants affect species composition. Most importantly, there is the danger that a new road will cause more fragmentation of ecosystems and contribute to the process of insularization.

Ecological evaluation and land use planning have been particularly well researched in the Netherlands, a country with one of the highest population densities in the world (330 per km², compared to the UK with 230 per km² or Japan with 292 per km²). Since the late 1960s, there have been a number of ecological evaluations undertaken in the Netherlands, many of which have made use of the ecotope concept; that is, vegetation-related data are collected on the basis of an ecotopic landscape pattern (p. 57).

In a review of 10 ecological evaluation methods used in the Netherlands (van der Ploeg and Vlijm, 1978), it emerged that most methods use data from plant species and plant communities. For example, the Landelijke Milieukarterung or environmental survey was based mainly on area, species richness and plant community richness. Ecotopes form the basis of the evaluation and the total value of the area is the sum of the values attributed to each ecotope as follows:

$$\text{Total value} = m\sqrt{\frac{\Sigma \, Ap_i \, w_i^m}{\Sigma \, Ap_i}}$$

where m is a general assessment coefficient (if m was 2, the higher component value contributes substantially more to the total value than the low component values, if m was infinity then the highest component value would determine the total value), Ap_i is the relative area of the ith ecotope, w_i is the value of the ith ecotope on a scale of 0–9. The value of each ecotope is determined mainly on the basis of nationally rare species and nationally rare plant communities.

Of particular interest in the research on ecological evaluation in the Netherlands is the concerted and valuable efforts to try and incorporate data on animal species richness and diversity. Whereas it is easy to record plant species richness and diversity (assuming accurate and reliable taxonomic information), many animal groups are difficult to locate and record. Some animal groups such as birds and butterflies are not difficult to record but at the same time it is important to note the temporal aspects of sampling (see section 39). Rarity, abundance and species richness (of taxa such as birds, amphibians, reptiles, lepidoptera and mammals) are the common criteria used for zoological evaluations. For example, in two projects, the Green Space Arnheim-Nijmegen (GRAN) project and the study for the regional planning area Midden-Gelderland (GRIM), zoological evaluations were calculated on the basis of:

$$\text{Total value (GRAN)} = \sum_{i=1}^{n} R_i \times T_i$$

where *R* is rarity of the species (birds, mammals, amphibians, reptiles) and *T* is the abundance of the species.

$$\text{Total value (GRIM)} = \sum_{i = 1}^{Nsp} \frac{T_i \text{ (local)}}{T_i \text{ (national)}} \times 100$$

where *T* is the abundance of the species (birds, amphibians, reptiles, mammals).

The data can then be expressed either in the form of maps similar to that used by Tubbs and Blackwood or in the form of lists of sites with indices used to express the value of the sites. A commonly used grading system for the maps has been as follows:

5. Very valuable with recommended designation of sites and complete protection supported by appropriate management and administration;
4. Valuable with recommended designation of the site and protection sufficient to allow only minor encroachments but in agreement with the conservation authorities;
3. Valuable but no recommended designation;
2. Some value but no recommendation;
1. Scant value with few precautions against encroachment.

Provided the underlying mathematics is explained, then indices can be a useful way of summarizing data. The alternative to an index of ecological value is to give a written account, usually with qualitative data. However, the quality of written accounts is very variable ranging from those based on casual observations of the more conspicuous flora and fauna to those which give detailed species lists based on extensive ecological surveys.

7.2.2 Views from planners and architects

Although ecologists, environmental scientists and conservationists have been the main proponents of evaluations and assessments, there have been a few planners and architects who have entered the debate about the methods to be used for assessments and evaluations of biotic communities. Working on London in the 1970s, Pickering (Maurice Pickering Associates, Architects and Environmental Planning and Landscape Consultants) was looking for a method of comparing the value of an ecosystem with the human impact on it and he suggested the principle that ecological value is inversely proportional to human impact (that is, the greater the disturbance and interference by humans, the less the value of the areas). From this he derived two formulae (Box 7.3), one for Ecological Value (EV) (reminiscent of Goldsmith's Index of Ecological Value (p. 122)) and one for Human

Box 7.3 The basis of Pickering's method of ecological evaluation for planning. The example here is for a part of London. Published with kind permission of Maurice E. Pickering.

Pickering's Principle

ECOLOGICAL VALUE IS INVERSELY PROPORTIONAL TO HUMAN IMPACT

so that

FOR ANY GIVEN LOCATION THE ECOLOGICAL VALUE (E) AND THE HUMAN IMPACT (H) MUST TEND TOWARDS EQUILIBRIUM.

The principle may be stated as:

$$E + H \longrightarrow 0.$$

where the index of Ecological Value,

$$E \equiv \log_{10} \sum_{1}^{n} \left[v^3.E.R.S. \right]$$

$$= (v_1^3 \times e_1 \times r_1 \times s_1) + (v_2^3 \times e_2 \times r_2 \times s_2) \ldots + (v_n^3 \times e_n \times r_n \times s_n)$$

in which:

v = number of vegetation horizons present
e = extent of habitat in hectares
r = rarity = $(100 - c)$, a percentage $c = \dfrac{e \text{ (extent of habitat) ha}}{A \text{ (Area of land system) km}^2}$

s = species diversity (number of species present)

and the sum of human events, Human Impact,

$$H \equiv \left[1.5 \log_{10} \sum_{1}^{n} \left(\frac{P.t \text{ av/day}}{2.5 \, z} \right) \right] - 10$$

in which:

P = number of persons in occupation
t = average duration of occupation in minutes per day
z = zone of occupation in hectares

(1) *v* equals number of vertical layers in the habitat. In Britain, this will not normally exceed four (tree, shrub, herb and ground). Thus, for a habitat with all four layers, *v* = 4; for a habitat with layers only to herb, *v* = 2.

(2) *r* equals the rarity factor, expressed as a percentage, which gives an indication of how common the particular habitat is in the land system under consideration. Formally, *r* can be expressed as:–

$$r = 100 \left(\frac{1 - \text{extent of the habitat in the whole land system}}{\text{area of land system}} \right)$$

But, in practice, *r* will normally be given a value as a result of estimation from a general knowledge of the land system rather than calculation.

Box 7.3 continued

(3) *s* equals the number of species (both vegetable and animal) on the site. Again, there is some estimation, with species of flora and fauna identified and counted but the number of insects deduced from a knowledge of the host plants.

(4) *P* equals the number of persons in occupation for a particular zone and for a particular activity. Vehicles count as the number of seats whether occupied or not (for instance, car = 4: coach = 60) with heavy lorries given the value of 60.

(5) *t* equals the duration of human occupation, expressed in minutes as the figure for the average day. Thus the time is measured for a particular activity during a normal day and then averaged by a figure for days/weeks or weeks/year if this activity only goes on for a certain number of days in the week or (as in school activity) weeks in the year.

(6) *z* equals zone of occupation in hectares. This may be varied to find the different values for *H* at varoius distances from the centre of activity being measured.

(7) *e* equals extent of habitat in hectares.

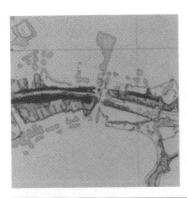

Box 7.3 continued

HUMAN ACTIVITY INCREASES

ECOLOGICAL VALUE INCREASES

	EV	*H*
Nature reserve (closed)	9·90	−10·00
Nature reserve (open)	9·80	−9·49
Forestry (closed)	8·20	−9·86
Forestry 9 (open)	8·10	−8·74
Agricultural land	7·56	−9·10
Public open space – woodland	6·70	−5·65
Public urban park (50 ha)	6·70	−5·60
Private campsite	6·65	−5·10
Playing fields	6·40	−6·40
Riding school	6·10	−5·95
Gravel workings (active)	5·60	−6·55
City open space – one ha	5·15	−3·85
Rural road	4·70	−4·50
Private garden	4·30	−5·05
Rural footpath	3·90	−4·38
Motorway	3·70	−3·40
Low density houses	2·00	−2·43
Urban fotpath	1·95	−1·68
High density houses	1·60	−1·45
Lorry park	1·50	−1·30
Public building	1·30	−1·15

impact (H). Pickering chose to cube *v* for two reasons: first, since the number of vegetation horizons will always be small (4–5), the contribution of this factor to *E* would be disproportionately low compared to e,r and s: second, he felt it right to give it more weight (M.E. Pickering personal communication). The number of species *s* is always a problem and in this case Pickering used trees, shrubs, herbs, birds and some other animal taxa so long as data collection did not impede a rapid response in the planning process.

For any given area, H and EV values are calculated on the basis of information from surveys and isopleths or contours are plotted on maps (Box 7.3). The self-checking effect of calculating an index from both formulae means that it is possible to calculate the value of H for a proposed development from accommodation schedules and directly infer the EV values (Pickering, 1977). This method, which was endorsed at the time by the local authority, can result in an objective assessment and contribute to rational planning decisions. Examples of values for EV and H published in a review of Pickering's method (Marsh, 1978) do show a surprising agreement between the two indices (Box 7.3).

Pickering remains convinced (personal communication) that a soundly argued ecological evaluation can assist considerably those assessments which in landscape terms may be wholly subjective and he also feels that ecological assessments must be seen to stand equal with human and economic factors. There always remains the formidable problem common to many methods of assessment and evaluation (but not the basic map exercises such as suggested by Tubbs and Blackwood) of the time required to collect biological and ecological data (note that Pickering relies mainly on botanical and selected zoological data, some inferred from knowledge about insect–plant relationships (Box 7.3)).

7.3 ENVIRONMENTAL IMPACT ASSESSMENTS (EIAs)

Environmental Impact Assessments (EIA) are systematic and formalized procedures for assessing the consequences of impacts (whether beneficial or damaging) arising from major projects and developments on various aspects of human welfare and all components of the environment. Particularly important is the function of EIAs in assessing whether or not various aspects of a development or project conform with statutory requirements or alternatively are perceived as being environmentally acceptable. If not acceptable then alternatives or modifications to various aspects of the development may have to be considered. Thus the objective of an Environmental Impact Assessment is to identify and quantify effects of primary, secondary and tertiary impacts (such as physical disturbance, pollution) on all aspects of the environment including fauna, flora and ecosystems. An Ecological Impact Assessment may be included in an EIA but would be concerned only with impacts on the ecology and ecosystems.

It is now nearly 30 years since EIAs were first developed as basic screening processes for environmental impacts arising from developments although general procedures for determining likely impacts have been undertaken over a much longer time. In the United States, the National Environment Policy Act (NEPA) passed by the United States Congress in 1969 required federal agencies to 'include in every recommendation or report on proposals for legislation and other major Federal actions significantly affecting the quality of the human environment a detailed statement . . . on the effects of that action.' It was that Act which first brought in a requirement for EIAs for federally funded projects in the USA.

Although the EIA process was developed in the 1960s many countries now have legislation requiring an environmental impact assessment prior to the commencement of any major development (detailed examples of legislation are given in Roberts and Roberts, 1984; Wathern, 1988 and Fortlage, 1990). The rationale underlying the formal (statutory) EIA is to make sure that the development will not affect human health and safety and that the development will not cause damage to the natural and physical environment. An EIA also provides a basis for looking at alternative sites, assessing the effects of different scales of developments and minimizing actions which may infringe on human health and safety.

The precise steps for an EIA are dependent on specific legislation but in general there is a sequence of steps commencing with a project proposal and screening; a decision is taken as to whether or not an EIA should be undertaken. There can then be a consultation phase. The consultation phase can be of considerable benefit to the developer, particularly with regard to the planning of the subsequent environmental assessments and in preparing a smooth passage for the whole environmental process. A project outline would usually incorporate information about the location and physical limits to the development. Whereas defining the physical limits of the site can be done objectively, defining the area affected by other impacts such as airborne effluents cannot be undertaken in a definitive manner without detailed and prolonged studies and so may not be undertaken at all.

A preliminary, outline EIA or scoping exercise, where technical details of the development are prepared together with preliminary environmental information, can be important for assessing the cost-effectiveness of full scale EIAs. The aim of the scoping exercise is to identify the scope and depth to which impact assessments should be undertaken and to focus attention on information gaps. Another important function of the preliminary EIA is to identify the expertise and personnel who will eventually be required to undertake the full environmental assessments.

Potential impacts may be easy to identify but then there is the more difficult task of assessing the magnitude of impacts on the various components of the environment. This part of the process is commonly and not surprisingly directed largely towards the human environment. Three basic methods for identifying impacts are: flow diagrams or networks (Figure 7.2), matrices, and checklists. Of those, checklists are seemingly the most simple, but various methods have been refined to quantify the impacts (see any of the following for examples: Munn, 1975; Canter, 1977; Jain *et al.*, 1977; Clark *et al.*, 1980, 1984; Black, 1981; Roberts and Roberts, 1984; Westman, 1985). Networks are particularly useful in that primary as well as secondary and tertiary impacts can be identified (Figure 7.2).

Other well-known methods for identification of impacts have been based on matrices, some of which can be used to give a subjective assessment of the magnitude and importance of the impact. The Leopold Matrix is perhaps the most well-known matrix which has been used in EIAs (Leopold *et al.*, 1971; Parker and Howard, 1977). In that matrix, environmental characteristics and project actions are recorded on an open matrix with up to 8800 interactions. First, the matrix is used to record those impacts (interactions) which are likely to occur and if the impact is 'significant' then a diagonal line is entered in a box. The interaction is then scored from 1 to 10 (1, low and 10, high) for both magnitude of the interaction (upper left) and the relative importance (lower right). Despite the criticisms that have been levelled at the Leopold Matrix (for example, it does not cater for secondary or tertiary interactions) this kind of approach has a cost-

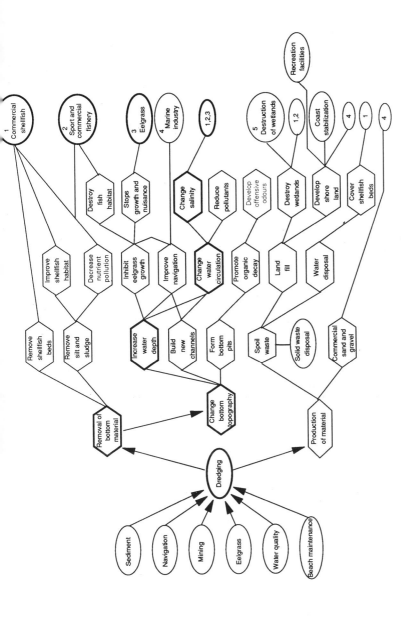

Figure 7.2 An example of a conceptual network for identifying impacts for an EIA on an estuary. In the boxes shown in bold, it can be seen that the indirect effects of dredging are varied. From Canter (1977).

effective role to play in providing a basis for monitoring changes in ecological systems.

7.4 ECOLOGICAL ASSESSMENTS

7.4.1 Concepts and methods

Gathering information on the likely effects of impacts on human communities and the built environment is an integral part of the EIA process. Most EIAs would usually include some information on potential impacts on the biology of the area: such as identification of protected areas and protected species within the area of potential impact arising from the project. Also, species of commercial interest and perhaps taxa of scientific interest may be the subject of study.

An Ecological Impact Assessment may or may not be included in an EIA and would be concerned with impacts on the natural environment, fauna and flora and ecological processes. An ecological assessment (compare with ecological evaluation) is used here to refer to investigations which determine the sensitivity of fauna, flora, habitats and ecosystems to physical or chemical impacts.

As with ecological evaluations, ecological assessments have to begin with surveys, classifications and inventories of natural areas and habitats. Identification and classification is very much open to subjective assessment and at the very least the recorder should prepare clear descriptions and definitions of the units used in surveys. In Britain, the publication of the first volumes of a standardized National Vegetation Classification (p. 54) will help to avoid much of the previously used subjective approaches to habitat classification.

Remote sensing, various computer programs and Geographical Information Systems (GIS) are helping to improve the accuracy and efficiency of environmental assessments, especially large scale assessments. For example, modelling the likely impacts of oil spills has become very sophisticated with the use of GIS. Large and complex data sets, such as maps of sensitive areas (sensitivity based on indicator species and communities) are easily stored, updated and communicated with modern information technology. In brief, when used as a basis for environmental sensitivity mapping, this modern technology allows the following (Jensen *et al.*, 1990):

1. Accurate base maps to be derived whenever a satellite image is available;
2. More accurate identification of oil-sensitive shoreline;
3. The use of a relational database GIS to store and query the information on oil-sensitive wildlife and access and protection.

In environmental assessments it is impractical to include information about all taxonomic groups or all species. Therefore, groups and species have to be selected with preferably a good rationale underlying the selection. Fry *et al.* (1986) have proposed useful guidelines for species selection in relation to developments within a particular region. The essence of the guidelines is a logical structure for selecting species as a basis for a database of wildlife habitat relationships. In outline, their method for species selection is as follows:

1. Identify all habitat types within the region;
2. Establish a database for species in the habitats and the use made of those habitats by the species (feeding areas, wintering areas etc.);
3. Screen the species for those which do not occur in the project area and for those species which have special habitat requirements not found in the project area;
4. Arrange the species in order of taxonomic class;
5. If appropriate some of the species could be grouped according to guilds;
6. Bearing in mind cost and time constraints, rationalize the lists of species to be used in the habitat evaluation. Determine how many species are required from each taxonomic class for optimal allocation of species then make the proportions among taxonomic classes in the sample equal to those in the candidate list;
7. Select individual species to fill the taxonomic class categories.

7.4.2 The nature of biology and ecology in EIA reports

With many EIA reports being undertaken, there have been opportunities to assess the biological and ecological aspects of those reports. In the absence of guidelines as to what biological and ecological aspects the EIAs should consist of, it is not surprising that there is wide variation and frequent misuse of ecological terminology. This was one of the conclusions reached after an investigation into the nature and use of biology and ecology in EIAs and environmental assessments undertaken in Britain during the period 1988–89 (the EEC Directive on 'the assessment of the effects of certain public and private projects on the environment' came into force in July 1988). A total of 45 EIA reports were examined and discussions were held with 26 organizations actively involved in EIA projects (Spellerberg and Minshull, 1990). Common to all the reports was a lack of distinction between ecology (concerned mainly with processes and interactions) and biology (largely species lists and distribution maps). In most cases where ecology was mentioned, the report was in fact dealing with biology. Of the 45 reports examined in detail, 84% included data gathered from original fieldwork but the objectives of collecting those data were not stated and the limitations of sampling methods were either not mentioned at all, or there was no attempt to quantify the limitations or express confidence limits. Also, few reports attempted to assess the validity of the results. Most of the reports (78%) included data from vegetation surveys and only 47% included information on animals (mostly birds).

7.4.3 Ecological assessment in practice

Ecological assessment methods have been used for classification of the level of sensitivity and vulnerability of coastlines to oil spills, sometimes culminating in coastal sensitivity maps. There are very good logistical reasons for undertaking this kind of assessment because at the time of a response to an oil spill or other pollution incident, those people who caused the incident will have an interest in undertaking cost-effective remedial measures. They will need to know where best to deploy the various kinds of recovery and clean-up operations. Classifying coastal regions from strict ecological points of view or via subjective assessments of sensitivity to water-borne oil has practical applications in not only contributing to better management of clean-up operations but also in siting installations.

The techniques for assessing and recording the sensitivity and vulnerability of coastal areas to pollution, especially oil spills, have been well researched. In North America, for example, work on ecological assessments on coastal areas has been undertaken by Hayes *et al.* (1980), Gundlach and Hayes (1978) and more recently by Jensen *et al.* (1990). Gundlach and Hayes developed a subjective scale of 1 to 10 based on type of coast (rocky, exposed headland, salt marsh) which is the basis for calculating a vulnerability index. In general, salt marshes and mangroves would be rated high in terms of the index whereas exposed headlands would be rated low. As more and more data are required for better and better assessments of sensitive coastal areas in relation to perturbations such as oil spills, simple distribution maps can no longer cope with the amount of data. However, it is now possible to store information about the spatial and temporal distribution of sensitive taxa in GIS (see p. 220) which not only allows rapid retrieval but also quite complex manipulation of large data sets (Jensen *et al.*, 1990).

Ecological assessments of coastlines in relation to the sensitivity of habitats and regions to water-borne oil have been undertaken in Britain in relation to antipollution contingency strategies. Examples have been undertaken around Southampton Water (Shears, 1990), at Sullom Voe (Figure 7.3) in the Shetlands (Syratt and Richardson, 1981; Shears, 1990) and also in the area around Poole Harbour (the second largest natural harbour in the world, second to Sydney Harbour), on the south coast of England (Pearson, 1984; Gray, 1985; BP Petroleum Development, 1986). The Sullum Voe site looks desolate and uninteresting (Figure 7.3) but nevertheless an extensive EIA and ecological assessment was undertaken in what is in fact an area rich in wildlife, some of considerable conservation interest (SOTEAG, 1989).

Figure 7.3 The Sullom Voe Oil Terminal, Calback Ness, Shetland looking west behind the main flare stack. Photograph kindly supplied by BP Co. plc 1192/6A.

The Institute of Terrestrial Ecology has carried out on behalf of the Nature Conservancy Council (see Box 6.1 for changes in status of the NCC) a survey of the Shetland coastline, analysing each point where it was intersected by the 1 km national grid. To ensure that priority is given to special features, as part of the oil spill contingency plans, the coast was assessed for the following features (Syratt and Richardson, 1981):

1. Ornithological interests
2. Mammals (including seals, otters and livestock)
3. Inshore fisheries
4. Salt marshes and soft mud shores
5. Houbs and other sensitive shallow water areas
6. Sites of Special Scientific Interest
7. Secondary beach structures such as spits and bars
8. Amenity beaches and other amenity areas

Along with is broad-scale assessment, sensitive areas were mapped, protective measures agreed and clean-up methods devised (Table 7.3).

In the south of England, a more detailed approach has been adopted in

Table 7.3 Oil spill response and fisheries information produced for the Sullom Voe Oil Spill Advisory Committee under the auspices of the Shetland Oil Terminal Environmental Advisory Committee for inclusion in the Sullom Voe Harbour Oil Spill Plan. Kindly provided by BP Exploration Operating Co. Ltd., Operator for Sullom Voe Terminal

Colour coding of coastline	Coastal features and biological areas	Comments	Protective measures	Clean-up methods			Protection and clean-up options
				Preferred	Possible	Avoid use of	
Red	Salt marshes Mud flats houbs	Sensitive areas. Avoid oil entry. Leave alone where possible. Gross contamination to be removed by hand tools. Generally of high biological interest. Physical disturbance or dispersant application may be damaging. Oil should be prevented from entering these areas. Recovery rate is slow.	1, 2	1, 6, 7	7, 8, 9, 12	4, 10, 11, 13	1. Containment and recovery using booms. 2. Direct to less sensitive shore to avoid widespread contamination. 3. Sand/shingle barriers. 4. Chemical dispersants. 5. Skimmers/ vacuum pumping. 6. Natural cleansing.
Yellow	Exposed coastline cliffs	Largely self cleaning due to high energy wave exposure. Access generally difficult. Safety considerations are essential. Recovery rate is moderate to fast.	1, 4	1, 6, 9	4, 5, 10	11	

Colour	Type	Description			
Blue	Accessible shoreline. Low lying shore Rock Cobble Shingle Gravel	Requires manual cleaning. Limiting factors will be weather, tides and available light. Manpower management and safety considerations are important. Operations should disturb shore structure as little as possible. Transportation and light mechanical cleaning equipment to have low pressure tyres. Recovery rate is moderate.	1, 2, 3	1, 5, 6, 7	4, 9, 12, 10, 11 13
Green	Amenity beaches Spur boom containment areas	Potential impact sites. Beaches can be cleaned with heavier mechanical plant if required. Recovery rate at spur boom sites is moderate. Sand beaches are low in ecological diversity.	1, 2, 3	1, 5, 7, 9	4, 8, 12, 6, 10, 13 11
Purple	Man-made structures, jetties, piers, slipways and gabions	Amenity structures. To be cleaned as a priority and as circumstances dictate.	1	1, 4, 5, 10, 11, 12	

7. Manual beach cleaning.
8. Sorption methods.
9. Low pressure sea water flushing.
10. High pressure flushing.
11. Hot water/steam hosing.
12. Vegetation removal.
13. Substrate removal manual mechanical.

Box 7.4 Environmental sensitivity map for part of Poole harbour, southern England (the original version is in colour). For the purposes of this environmental mapping, Poole harbour was divided into 13 conveniently sized areas and this is one example. The assessment was undertaken in relation to a subjective assessment of the sensitivity to water-borne oil and based largely on intertidal flora and some birds. The bird data were pooled into three main groups, representing intertidal species, gulls and diving birds, all of which use Poole harbour in different ways. In each case the area of a sector in the pie chart represents the number of birds in that group which might be found in that section of the harbour in a particular month (based on data using the highest 10 counts over a previous census period between 1969 and 1983). Modified from Gray (1985) with kind permission of the author and NERC.

North Arne

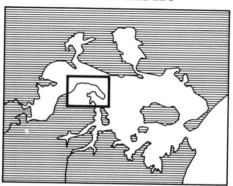

The two sides of the northern tip of the Arne peninsula provide a contrast in sediment and shoreline types. Along the west and north shores sand and gravel beaches, backed by cliffs and wind-blown sand, reveal a high energy shoreline. Beach material from the cliffs has been carried eastwards beyond Gold Point, where it protrudes as a cuspate foreland, to form a long spit, recurving at its tip (Patchins Point) and within which marshland developed before the advent of *Spartina*.

In contrast, on the eastern shore are the sheltered tree-lined marshes and mudflats of Arne Bay. Protected by Patchins Point to the north and a spit which has grown northwards from Shipstal Point cliff in the south, the Bay was colonized by *Spartina* around 1900 and became an important area for the export of the grass. Largely through the efforts of Mr R Cartridge, around 175,000 *Spartina* fragments and many seed samples were exported from Poole Harbour between 1924 and 1936 alone; Arne Bay being a key area (Hubbard 1965). Exported to more than 130 sites around the world, the grass was used for land reclamation and sea defence programmes. Today the area of *Spartina* in the Bay has declined and much of the lower marsh has an erosion cliff at the front of the sward. Despite this, the sediments to seaward of the marshes, and of the steeply shelving sandy beaches along the north shore, consist largely of very soft muds.

Box 7.4 continued

The shorelines are generally remote from public access from the land, and form part of the Arne Nature Reserve of the Royal Society for the Protection of Birds. Added to this the variety of shoreline types and near-shore habitats, ensures a high diversity of bird species. The most numerous wader is the oystercatcher, at times the area holding more than half the Harbour total, but species as different in their requirements as turnstone and curlew occur in relatively high numbers. The Arne Bay-Patchins Point area is important for shelduck and other waterfowl and the water to the east contains good populations of diving birds and a night-time gull roost.

ECOLOGICAL SENSITIVITY

Birds: Abundance
and seasonal
variation

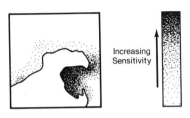

Increasing
Sensitivity

LAND USE

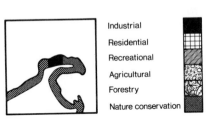

Industrial
Residential
Recreational
Agricultural
Forestry
Nature conservation

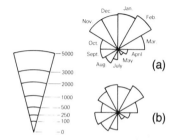

(a)

(b)

(c)

Birds: Seasonal
variation

(a) INTERTIDAL

The variety of shoreline types is reflected in the numbers and diversity of species counted in this area. The most numerous intertidal species is the oystercatcher; the average high count of 555 representing around 54% of the Poole Harbour total. The highest numbers generally occur in Arne Bay but oystercatchers are well scattered around the area. In contrast, almost all of the turnstone (14, equivalent to 35% of the Harbour total), ringed

Box 7.4 continued

plover (29 ≡ 20%) and grey plover (13 ≡ 8%) have been recorded along the north shore east of Gold Point, and all of the black-tailed godwit (51 ≡ 9%) within Arne Bay. Curlew (267 ≡ 25%) are rather more scattered but mainly occur in the Patchins Point area, whereas greenshank (14 ≡ 36%) are most common on the north shore west of Gold Point. Of the ducks, shelduck (404 ≡ 17%), wigeon (215 ≡ 12%) and teal (139 ≡ 10%) occur mainly in the Arne Bay-Patchins Point area.

(b) GULLS

Gull numbers are relatively low on this section of shoreline, the most notable feature being the relatively high numbers of herring gull. However, a large night-time gull roost (c. 5000 black-headed gulls) was observed in November 1984 on the open water west of Arne extending towards Brownsea Island.

(c) DIVING BIRDS

Diving birds occur offshore in small, but significant numbers, with both cormorant (46) and red-breasted merganser (60) reaching at times around a quarter of the Harbour average total. The area also holds proportionally high numbers of great-crested grebe and goldeneye.

---

an effort to identify sensitive coastal areas. Each section of the shoreline in Poole harbour was assessed in terms of its relative sensitivity to damage by water-borne oil (Box 7.4). Indicator species of some intertidal plants and animals, particularly birds, provided the basis for the composite shoreline 'ecological' sensitivity map. The limitations of this sensitivity assessment (as discussed in the publication) include the following: the assessment does not take full account of the interdependence of different parts of the harbour ecosystem, intertidal ecosystems are dynamic and therefore such an analysis is essentially a 'snap-shot' of present conditions; and assessment of ecological sensitivity specifically ignores questions of land use or evaluation. However, despite these and other limitations, sensitivity maps of this kind form the basis for pollution response contingency plans.

7.4.4 An operational framework for an ecological assessment

Although many countries now have their own statutory requirements and guidelines (statutory instruments) for environmental assessments of the living environment, those requirements and guidelines are far from precise and sometimes not informative. Clearly there seems to be much room for improvement and more research and more discussion needs to take place. As a contribution to that discussion the following outline is given as a basis for an environmental assessment of the biology and ecology of a hypothetical estuary.

Any assessment of the biology of the estuary would start with questions of what to record and how to record. At one level the assessment could be based on habitats (see p. 55), for example, salt marshes, rocky shores and the littoral zone. Particular attention needs to be paid to protected areas of which there may be many kinds including nationally designated marine nature reserves and sites of special scientific interest as well as sites of international importance such as Ramsar sites. Information about the presence of protected areas is provided by local government and by conservation authorities.

In view of the impractical nature of surveying all taxonomic groups in an estuary, there has to be selection. That selection might include the following groups in no order of importance:

1. Protected species: those species protected by law;
2. Easily recorded taxonomic groups such as birds, butterflies and flowering plants;
3. Sensitive species: these could be indicator species and species sensitive to various kinds of disturbance and pollution;
4. Species on red data lists;
5. Commercially important species such as shellfish;
6. Species perceived as being attractive by the local community, for example swans.

When and how the area is surveyed is very important because different species will occur at different times. Furthermore, it should always be borne in mind that all survey methods have limitations and may be biassed towards certain species (p. 59).

Environmental impact assessments often require information on the distribution and abundance of those species which may be affected by any of the impacts (primary, secondary, tertiary) arising from the development of the project. The simple use of checklists and matrices can be used to make those identifications but networks are particularly useful for identifying secondary and tertiary impacts (Figure 7.2). Those species of particular interest for assessing impacts and for use in identifying sensitive areas are those which have narrow tolerances to physical and chemical conditions. There is a growing accumulation of data on the tolerances of marine organisms, especially estuarine species: data may be deposited in marine research laboratories or may be found in reference publications.

Establishing definitive species lists may be impractical and obtaining fully comprehensive information on the ecology of an estuary is virtually impossible. This is not to say that detailed information on certain aspects of the ecology could not be obtained within a short time. The potential range of ecological topics can be considerably varied (Table 7.4) and clearly there has to be a selection of topics. Biomass, productivity, and population structure of selected taxonomic groups would be particularly important for environmental impact assessments. Other aspects of the ecology included in an ecological assessment would need be only those which are directly relevant to the conservation aims and human use of the region.

Table 7.4 Some potential ecological components of EIAs in estuaries. A list of examples of ecological processes and variables which could be investigated as part of an EIA in an estuary. This list is by no means complete but serves to show the breadth of potential ecological components

---

Effects on Community structure as measured by changes in
    Species composition
    Species richness
    Species diversity
    Biotic indices

Effects on the following (in relation to specific taxonomic groups)
    Standing crop or biomass
    Productivity
    Production
    Population density
    Reproduction
    Breeding sites
    Migration
    Dispersal
    Habitat carrying capacity

The following is a basis for an operational framework for the biological and ecological component of an EIA:

1. Identify the limits to the area;
2. Obtain information about extent and kind of protected areas;
3. Select a habitat classification and record types of habitats;
4. Select taxonomic groups to be surveyed and appropriate data collection methods;
5. Identify primary, secondary and tertiary impacts and assess the nature and magnitude of the potential impacts;
6. Design and implement monitoring procedures.

This is a basic operational framework which oversimplifies real circumstances, especially with regard to defining the spatial and temporal limits of the impacts. For example, even if EIAs had been compulsory 20 years ago for coal-fired electricity-generating stations, it is doubtful if EIAs would have predicted that the gases would affect habitats as far away as Scandinavia and Europe. The effects of 'acid rain' are now well known and there seems to be fairly general agreement that electricity-generating stations have contributed on a large scale to this environmental problem. Although there has been a cumulative effect in terms of the 'acid rain', it would have been a far-sighted person who could have predicted such distant effects (in space and in time) of proposed coal-fired electricity-generating stations on the forests and lakes of Scandinavia and Europe.

7.5 CONCLUSIONS

At least four important points emerge from this brief survey of methods for ecological evaluation and ecological assessments in relation to planning and environmental impact assessments: first, the need to appreciate the implications of change in natural systems; second, the need to quantify data where possible; third, the reliability of the data should be made known; and fourth, there should be a way of validating taxonomic information.

Many environmental assessments in relation to planning and development have been designed and restricted to snapshots in time. These assessments require a lot of effort and until recently there has been no way of regularly updating the assessments, which is a pity because environments are always changing. The question, at what frequency need the evaluations or assessments be updated?, is one which can only be answered with information about ecosystem processes. If we are really concerned about the effects of perturbations on natural communities, then it seems important in environmental assessments to include data on ecological processes as well as ecological patterns. That is, it is important to know not only what is there but also something about ecosystem properties. Obviously much more information needs to be collected but with remote sensing, GIS and advances in information technology there seems to be every opportunity to explore ways of incorporating the temporal aspects as well as the spatial aspects of an ecological process. Furthermore, it would seem possible not only to assess habitat sensitivity to certain perturbations but also the ability to recover. By ability I mean the time taken to recover to certain levels of community structure and whether or not restoration and management would be necessary to help the process of recovery.

Research on methods for incorporating ecosystem properties into management planning has been undertaken for some years in North America, especially in connection with wetland ecosystems such as around Green Bay, Lake Michigan (Figure 5.5). There, it has been recognized that evaluations and assessments must be monitored and that it is of little value to plan a 'one-off' assessment (Harris *et al.*, 1981, 1987). At the same time, it has been argued that a set of ecosystem properties (see for example Table 7.4) is needed for assessment of conditions.

There has been much thought and research on how best to evaluate natural areas in relation to planning and environmental assessments; some methods being quite simple and others complex and culminating in an index. Despite the many attempts to quantify the conservation interest of an area it seems that in connection with strategic planning or environmental assessments, qualitative evaluations are preferred. However, qualitative surveys tend to be interesting not as a result of what they mention but what they do not mention. Therefore, conclusions based on qualitative surveys need to take into account what is not mentioned and why. Also very popular are evaluation maps showing the relative values of different areas. However, there seems to be growing concern that evaluation maps can be misused because of the relative values shown on the maps. It seems that areas of 'low' conservation value may sometimes be considered as areas where development is more likely to be agreed. In some countries such as the Netherlands, there is growing opposition to the use of land evaluation maps because of the possibility that such maps may be misinterpreted. Evaluation maps can have a valuable role to play in strategic planning but it must be recognized that they show only relative values of conservation interest on a broad scale. Further interpretation of the conservation value of the areas on the map require more detailed ecological evaluations where like is compared with like.

The value of ecological impact assessments can sometimes be limited, not by what is included in the results but by what is omitted. For example, descriptions of the sampling and recording methods are sometimes not sufficiently detailed for the reliability of the data to be assessed. There are well-tested ecological survey methods which ensure that there is species representative sampling of habitats of ecosystems and there are also well tested methods for assessing the validity of the sampling or survey methods. There are also many statistical methods for assessing the reliability of the data. If such assessments of the reliability of the data and the limitations of the methods are not described, then there should be an explanation.

If quantitative evaluations and assessments can be based on rigorous methods, then why the apparent preference for qualitative methods? One obvious reason is simply costs in terms of time and money. In general, qualitative surveys and general descriptions can be achieved more cheaply than the detailed quantitative approaches. This being the case and assuming that there is a preference for objective and reliable methods, perhaps there is a need for more research on cost-effective yet quantitative methods of evaluation and assessment of natural communities.

Many ecological assessments are based on the assumption that plants and animals have been identified correctly. There is an ever-increasing demand for identification skills and although there are many 'experts', it seems odd that there has been little effort to confirm the validity of identifications. We all know that we make mistakes and therefore if decisions are to be made on the basis of species lists, then it would seem only right to include confirmation of species identification in the process of ecological evaluation or EIA.

In theory, all natural and semi-natural areas should be conserved and all should be protected from changes in land use, especially if new projects threaten to reduce and fragment the area (the greatest threat to biodiversity: p. 12). In practice, evaluations and assessments are a necessity because of the many demands on the environment. But, what is the best method of evaluation and method of assessment for use in connection with planning and EIAs? Of the various methods described here, Pickering's method, although innovative, has too many stages and is based on too many assumptions. For use in connection with planning and identification of areas of conservation interest on a fairly large-scale (counties, provinces or states) the method developed by Tubbs and later adapted by the Dorset Trust for Conservation of Nature provides a good basic approach which could easily be modified for specific countries. On a smaller scale of around a few hectares or less, evaluation of areas (ranging from natural habitats to derelict urban sites) could usefully be undertaken using some of Ratcliffe's criteria (p. 186) with a few modifications. Qualification of rarity for example could be achieved with reference to international, national and regional status (see evaluation of limestone pavements on p. 123). Endangered status could be qualified using threat numbers.

Ecological assessments based on indicator species or sensitive species seem to have been fairly well received although there has been a criticism that insufficient quantification is incorporated. There is one group of projects for which more research is required if useful ecological assessments are to be undertaken in the future and they are linear projects such as construction of roads and railway lines. All too often an environmental consultancy is asked to undertake work on the potential impacts of the road after the route has been chosen: ecologists should be involved throughout, from planning to monitoring. There are two aspects in need of research; one concerns methods of incorporating costs into the project which represent losses of natural areas, and the other concerns methods which help to select routes with least damage to natural areas.

The planning of routes for roads and railway lines has for far too long disregarded costs of losses and damage to natural areas and especially protected areas; it is a topic in much need of more research because it is unacceptable that such losses are not included in the costs. The other aspect (selecting the least damaging to the natural environment) is intriguing because no single method has yet been adopted and there are a number of false assumptions as to how to proceed. A common kind of question when choosing the route is: 'how do you compare an internationally important wetland with an ancient woodland?' The answer is: you don't. You compare like with like as a basis for establishing priorities.

In choosing a route for a new road (in terms of maintaining conservation of biodiversity), designated protected areas must continue to be protected and should receive greater protection than conservation of human artefacts or buildings (manufactured objects can be rebuilt but generally speaking nature cannot be restored to what it was). The next step should be either to calculate ecological damage (perhaps on the basis of Yapp's method) or to undertake an ecological evaluation of each habitat (see Box 8.1 and Table 8.1). The road may have to go through some areas of relatively high conservation value but with systematic evaluations it should be possible to identify the least-damaging routes. Three other aspects are worth considering. Further fragmentation of habitats must be avoided or overcome in some way (perhaps with artificial corridors, such as badger tunnels, to allow dispersion between habitats). The impacts of roads does not stop a few metres from the edge of the road. It may be necessary to incorporate some kind of ecological buffer zone between the new road and some kinds of habitats so that the 'ecological integrity' of the habitats is maintained. Finally, if habitats are damaged or destroyed then there should be an accepted practice that similar habitats are created nearby: there should be no net loss of nature.

REFERENCES

Black, P.E. (1981) *Environmental Impact Analysis*. Praeger, New York, London.

BP Petroleum Development Ltd (1986) *Wytch Farm Development, Purbeck – Southampton Pipeline Environmental Impact Assessment*. BP Petroleum Development Ltd.

Canter, L.W. (1977) *Environmental Impact Assessment*, McGraw-Hill, London, New York.

Clark, B.D., Bisset, R. and Wathern, P. (1980) *Environmental Impact Assessment*. Mansell, London.

Clark, B.D., Gilad, A., Bisset, R. and Tomlinson, P. (1984) *Perspectives on Environmental Impact Assessments*. D. Reidel, Dordrecht, Boston.

Fry, M.E., Risser, R.J., Stubbs, H.A. and Leighton, J.P. (1986) Species selection for habitat-evaluation procedures, in *Wildlife 2000 Modeling Habitat Relationships of Terrestrial Vertebrates* (eds J. Verner, M.L. Morison and C.J. Ralph) University of Wisconsin Press, Wisconsin, pp. 105–13.

Fortlage, C.A. (1990) *Environmental Assessment: a Practical Guide*. Gower, Aldershot.

Gray, A.J. (1985) *Poole Harbour: Ecological Sensitivity Analysis of the Shoreline*. NERC Institute of Terrestrial Ecology, Huntingdon.

Gundlach, E.R. and Hayes, M.O. (1978) Vulnerability of coastal environments to oil spill impacts. *Marine Technology Society Journal*, **12**, 18–27.

Harris, H.J., Fewless, G., Milligan, M. and Johnson, W. (1981) Recovery processes and habitat quality in a freshwater coastal marsh following a natural disturbance, in *Selected Proceedings of the Midwest Conference on Wetland Values and Management* (ed. B. Richardson) St Paul, Minnesota, pp. 363–79.

Harris, H.J., Richman, S., Harris, V.A. and Yarbrough, C.J. (1987) Coupling ecosystem science with management: a Great lakes perspective from Green Bay, Lake Michigan, USA. *Environmental Management*, 11, 619–25.

Hayes, M.O., Gundlach, E.R. and Getter, C.D. (1980) Sensitivity ranking of energy port shorelines, in *Proceedings of Ports* (eds M.O. Hayes *et al.*) American Society of Civil Engineers, Norfolk, VA, pp. 699–709.

Jain, R.K., Urban, L.V. and Stacey, G.S. (1977) *Environmental Impact Analysis*. Van Nostrand Reinhold, New York.

Jensen, J.R., Ramsey, E.W., Holmes, J.M., Michel, J.E., Savitsky, B. and Davis, B.A. (1990) Environmental sensitivity index (ESI) mapping for oil spills using remote sensing and geographic information system technology. *International Journal of Geographical Information Systems*, 4, 181–201.

Leopold, L.B., Clarke, F.E., Henshaw, B.B. and Balsley, J.R. (1971) A procedure for evaluating environmental impact. *Circular, US Geological Survey*, p. 645.

Marsh, P. (1978) Formula for the needs of man and nature. *New Scientist*, 77, 84–6.

Munn, R.E. (1975) Environmental impact assessment: principles and procedures. *SCOPE Report*, 5, UNEP, UNESCO, Toronto, Canada.

Parker, B.C. and Howard, R.V. (1977) The first environmental impact monitoring and assessment in Antarctica. The dry valley drilling project. *Biological Conservation*, 12, 163–77.

Pearson, N. Associates (1984) *Wytch Farm Oilfield Development, Furzey Island Visual Impact Analysis*, BP Development Ltd.

Pickering, M.E. (1977) A new environmental planning tool. Building Technology and Management. *Journal of the Institute of Building*, July/August.

Ploeg, S.W.F. van der and Vlijm, L. (1978) Ecological evaluation, nature conservation and land use planning with particular reference to methods used in the Netherlands. *Biological Conservation*, 14, 197–221.

Roberts, R.D. and Roberts, T.M. (1984) *Planning and Ecology*. Chapman and Hall, London.

Rutter, A.J. and Thompson, N.E. (1986) The salinity of motorway soils III. Simulation of the effects of salt usage and rainfall on sodium and chloride concentrations in the soil of central reserves. *J. Appl. Ecol.*, 23, 281–9.

Shears, J.R. (1990) The environmental assessment of oil pollution. PhD Thesis, Southampton University.

SOTEAG (1989) Environmental monitoring at the Sullom Voe development in Shetland. A report compiled by members of SOTEAG. Shetland, The Shetland Oil Terminal Environmental Advisory Group.

Spellerberg, I.F. and Gaywood, M.J. (1993) Linear features: wildlife corridors and linear habitats. Peterborough, English Nature.

Spellerberg, I.F. and Minshull, A. (1990) An investigation into the nature and use of ecology in environmental impact assessments. Unpublished report to the British Ecological Society, Ecological Affairs Committee.

Syratt, W.J. and Richardson, M.G. (1981) Anti-pollution strategy in Sullom Voe – environmental considerations. *Proceedings of the Royal Society of Edinburgh*, 80B, 33–51.

Tubbs, C.R. and Blackwood, J.W. (1971) Ecological evaluation of land for planning purposes. *Biological Conservation*, 3, 169–72.

Wathern, P. (1988) *Environmental Impact Assessment. Theory and Practice.* Unwin Hyman, London.

Westman, W.E. (1985) *Ecology: Impact assessment and environmental planning.* John Wiley, Chichester.

Wise, A.J. (1977) *Wildlife conservation in the Isle of Purbeck. Dorset Naturalists' Trust Conservation Studies* 4.

Yapp, W.B. (1973) Ecological evaluation of a linear landscape. *Biological Conservation,* 5, 45–7.

–8

Epilogue

In the 1970s and early 1980s, many methods of ecological evaluation were proposed and described and they were soon followed by many reviews. During the early 1980s it seemed that there would be few new developments and initiatives of any theoretical or practical significance. I think that many ecologists and conservationists interested in evaluation believed that all that could be done with red lists had been done, that ecological evaluation had been well discussed and that ecological assessments of communities for use in planning had been researched to the extent that further research was unnecessary. On the contrary, there have been major advances in ecological evaluation and there are now many methods suitable for a variety of objectives (Table 8.1). In general there has been research and development towards more refined methods for quantification of the ecological criteria and thus a move away from the use of subjective assessments. There have been some excellent theoretical innovations in evaluation and assessment for nature conservation, although not all have been tested in practice. Examples of innovative developments in ecological evaluation and assessment include the work by Vane-Wright *et al.* (1991) on taxonomy and systematics (p. 82), the research on red data book categories by Mace and Lande (1991) (p. 91), and the work by Millsap *et al.* (1990) on species ranking systems (p. 104). Common to these new developments has been the attempt to use criteria which can be quantified. This is not to say ecological criteria only should be used. On the contrary, I believe that it is important to make use of a range of criteria, both ecological and non-ecological. Where ecology is used, it should be based on standardized, quantifiable and repeatable methods. That aspect is emphasized in the conceptual model shown in Box 8.1.

Box 8.1 Conceptual plan for an ecological evaluation.

Conceptual plan for an ecological evaluation

1) Define the objective of the evaluation.

2) Identify protected areas/species.

3a) Select potential methods of evaluation which are appropriate for the objective (see Table 8.1).

3b) Select classification recording and sampling methods appropriate to the scale and to the types of taxonomic groups. See for example Box 2.1, Fig 2.1, Fig 2.2.

4a) Undertake preliminary trial with selected method in a limited area of the locality to be evaluated.

4b) Test recording and sampling method and device appropriate tests of reliability. Arrange for confirmation of species identification.

5a) Modify the evaluation method (but justify).

5b) Modify the methods or design new methods (but justify).

6) Undertake the evaluation. Note any variable which could affect the results (e.g. time of data collection, type of survey and recording methods).

7) Analyse results giving qualifying information such as potential range of values and variance of any quantitative data.

This is not to say that there are no further questions or problems. One area urgently in need of continued research by sharp minds lies in the field of economics, ecology and philosophy. How best to approach the valuing of species and natural areas as a component of economic valuation of land is a difficult question to answer. This is not a new observation but it is a pertinent observation at a time when there is a lot of discussion (at the international and governmental level) about calculating the value of organisms, the natural environment and the ecosystems of the world. The greening of the economy, say some, is just around the corner but it seems to me that we are faced with some very difficult questions which will take some time yet to research. It has often been said that economists and ecologists should get together; that has been tried with little in the way of productive results. Perhaps the difficulties may partly be attributed to over-specialized training and therefore one way forward could be to establish some carefully designed joint (interdisciplinary) courses and meetings incorporating appropriate aspects of ecology, economics and philosophy. Interdisciplinary education seems to be attracting attention. For example, an interdisciplinary

Table 8.1 Examples of (a) ecological evaluation methods and (b) ecological assessment methods with their potential applications

Authority or author	Reference in this book	Application
(a) *Ecological evaluation*		
Tubbs and Blackwood (1971)	Section 7.1	Regional evaluation of semi-natural and agricultural ecosystems
Ratcliffe (1977)	Section 6.6	Designation of protected areas but also has potential for general evaluation and priority ranking
Ogle (1981)	Table 5.2	Temperate forests (New Zealand)
Peterken (1974)	Section 5.5	Temperate woodlands (Britain)
Kent and Smart (1981)	Section 5.2	Semi-natural vegetation (temperate regions) with particular reference to fragmentation
Goldsmith (1975)	Section 5.2	Agricultural and semi-natural habitats
Lancia *et al.* (1982)	Section 5.2	Habitat Evaluation Procedure: key habitat factors used to evaluate areas for certain species
Götmark *et al.* (1986)	Section 5.4	Avian habitats
Ramsar convention	Box 5.3	Wetlands of importance to birds
Harris *et al.* (1983)	Section 5.7	Wetlands
Morgan (1982)	Section 5.7	Inland waters
O'Keeffe *et al.* (1987)	Section 5.7	Rivers
Wittig and Schreiber (1983)	Section 5.6	Urban habitats
Ward and Evans (1976)	Section 5.2	Limestone pavements (Britain)
(b) *Ecological assessment*		
Yapp (1973)	Table 7.2	Ecological damage of linear features such as roads
Syratt and Richardson (1981)	Table 7.3	Ecological assessment of marine and coastal habitats
Gray (1985)	Box 7.4	Ecological assessment of coastal habitats

approach to teaching conservation biology seems to be popular in North America. Perhaps this is the way forward if we are to train people to deal with problems of conserving biodiversity in the future. True, conservation biology has existed as a discipline for a very long time but it has not, until recently, been taught as an interdisciplinary subject embracing taxonomy, systematics, population biology, ecology, genetics, biogeography, environmental economics and environmental education. I think centres of higher education throughout the world would do well to consider developing or strengthening such interdisciplinary courses.

Ecological evaluation and assessment is dependent on the availability of good data and also reliable identification of taxa. There is a need for the development of existing biological databases and research into how the information can be distributed effectively. In many countries there are a lot of data stored in many places in a variety of systems but access to those data is difficult. The quality and usefulness of many ecological evaluations and ecological assessments could be much improved if biological and ecological data were more easily accessible. The importance of systematics and taxonomy to conservation has been stated many times in this book; there is a need for more training and a need for standardized classifications. In relation to this, it is not before time that we are beginning to see new initiatives for the establishment of centres for biodiversity research.

While researching material for this book I have been in communication with many colleagues throughout the world and I have been lucky enough to acquire the unselfish help of many new colleagues. In doing so, it became very clear that there is yet much to be learnt from each other. A more international approach to the development of skills and research for evaluation and assessment could make a valuable contribution to conserving the world's biodiversity. But collaboration and communication is just as important at a much lower level. It has also become very obvious that on the one hand there is much exciting research and on the other only a small fraction of that research is being put into practice. There seem to be two reasons for this, first, the less than satisfactory means of communication between researchers and practitoners and second, the low level of recognition of the value and benefits of nature – which brings me back full circle to a theme in the opening chapter of this book. True, there is a growing awareness of the importance of conserving biodiversity but that awareness is not spreading among decision makers and into economics and politics as rapidly as it should.

It is thought that about 4% of the world's land surface has been designated as protected areas and there are now many forms of wildlife law. The world's protected areas together with wildlife law are not enough to balance the losses of biodiversity. As well as more protected areas, we need to conserve as much as possible between those protected areas. Any project (road, airfield, factory, housing estate etc.) that may damage or destroy natural or semi-natural areas should not proceed unless it can be shown (by the developer) that there is no other alternative. Claims that the costs of alternatives and the costs of delays are prohibitive will of course be made, but ecologists and conservationists must strive to discuss the importance of a cost–benefit analysis which takes into consideration the value of nature, the cost of avoiding a net loss in nature and the costs to the natural environment in the long term. This should be done by way of constructive discussion rather than confrontation.

Finally, the earth as our one and only home with its finite resources seems to be much valued when viewed from space. Many an astronaut has been deeply moved at their first sight of the earth from a spacecraft. How ironic therefore to hear about ideas and plans to create an atmosphere on other planets such as Mars, possibly making that planet fit for human habitation. Have we not already failed to care for one planet and have we not yet to find out a lot more about the Earth's biodiversity and what it has to offer? More and more animals, plants and landscapes are destined to survive only in memories and in pictures (Figure 8.1) unless conservation of biodiversity is undertaken by many more people at a much larger scale. Environmental degradation could be eased and the quality of life of millions of people for generations to come would be improved if conservation of biodiversity was accepted as being the most important environmental issue and financed accordingly.

Figure 8.1 Does it matter if a landscape is damaged if landscapes can be preserved for posterity as a painting and thus part of the national heritage? Original paintings by Jane Perry.

Appendix A: Further reading and reference material

Cutrera, A. 1991) *European Environmental Yearbook*. 2nd edn. DocTer International UK, London.

de Groot, R.S. (in press) Functions of Nature. *Description and evaluation of the functions of nature as a tool in environmental planning, management and decision making*. Wolters-Nourdhoff, Groningen, The Netherlands.

Helliwell, D.R. (1985) *Planning for Nature Conservation*. Packard Publishing, Chichester.

McNeely, J.A. (1988) *Economics and Biological Diversity*. IUCN, Gland, Switzerland.

McNeely, J.A., Miller, K.R., Reid, W.V., Mittermeier, R.A. and Werner, T.B. (1990) *Conserving the World's Biological Diversity*. IUCN, World Resources Institute, Conservation International, WWF and the World Bank, Gland.

Myers, N. (1983) A priority-ranking strategy for threatened species. *The Environmentalist*, **3**, 97–120.

Norton, B.G. (ed.) (1986) *The Preservation of Species: The value of biological diversity*. Princeton University Press, Princeton, NJ.

Norton, B.G. (1987) *Why Preserve Natural Variety?* Princeton University Press, Princeton, NJ.

O'Connor, K.F., Overmars, F.B. and Ralston, M.M. (1990) *Land Evaluation for Nature Conservation*. Department of Conservation, Wellington.

Shaw, W.W. and Zube, E.H. (1980) Wildlife values. Centre for Assessment of Noncommodity Natural Resource Values, *Institutional Series Report* No. 1.

Spellerberg, I.F. (1981) *Ecological Evaluation for Conservation*. Edward Arnold, London.

Spellerberg, I.F. (1991) *Monitoring Ecological Change*. Cambridge University Press, Cambridge.

Usher, M.B. (1986) *Wildlife Conservation Evaluation*, Chapman and Hall, London.

Van DeVeer, D. and Pierce, C. (1986) *People, Penguins and Plastic Trees*. Wadsworth Publishing Co., Belmont, CA.

Worster, D. (1977) *Nature's Economy. A History of Ecological Ideas*. Cambridge University Press, Cambridge.

Appendix B: Organizations mentioned in the text

Australian Conservation Foundation
340 Gore Street,
Fitzroy,
Victoria,
Australia 3065

Australian Heritage Commission,
GPO Box 1567,
Canberra ACT,
Australia, 2601

Australian National Parks and Wildlife Service,
GPO Box 636,
Canberra,
ACT,
Australia 2601

Berne Convention Secretariat,
Conseil de l'Europe,
F67000,
Strasbourg,
France

Botanical Society of the British Isles (BSBI),
Department of Botany,
British Museum (Natural History), Cromwell Rd,
London SW7 6BD, UK

British Ecological Society (BES),
26 Blades Court,
Deodar Road,
Putney,
London,
SW15 2NU

British Trust for Ornithology (BTO),
The Nunnery,
Nunnery Place,
Thetford,
Norfolk
IP24 2BR, UK

Centre for Marine Conservation,
1725 DeSales St NW,
Washington DC,
USA

Convention on International Trade in Endangered Species,
(CITES) Secretariat,
6 Rue Maupas,
Case Postale 78,
CH–1,000,
Lausane 9,
Switzerland

Council of Europe,
BP 431/R6,
67006 Strasbourg Cedex,
France

Council on Environmental Quality,
722 Jackson Place, NW,
Washington, DC 20503,
USA

Countryside Commission,
John Dower House,
Crescent Place,
Cheltenham,
Glos. GL50 3RA,
UK

Ecological Society of America,
Public Affairs Office,
2010 Massachusetts Ave., NW,
Washington DC 20036
USA

English Nature (Nature Conservancy Council for England),
Northminster House,
Peterborough PE1 1UA,
UK

Food and Agricultural Organization (FAO),
Viale delle Terme di Caracalla,
Rome,
Italy

Freshwater Biological Association (FBA),
The Ferry House,
Far Sawrey,
Ambleside,
Cumbria LA22 0LP,
UK

Institute of Freshwater Ecology,
Windemere Laboratory,
The Ferry House,
Ambleside,
Cumbria LA22 0LP,
UK

The Institute of Terrestrial Ecology (ITE),
Bush Estate,
Penicuik,
Midlothian EH26 0QB,
UK

International Board for Plant Genetic Resources (IBPGR),
Via delle Sette Chiese 142,
00145 Rome,
Italy

International Centre for Protected Landscapes,
Brecon Beacons National Park,
7 Glamorgan St,
GB Brecon,
Powys LD3 7DP, UK

International Council for Bird Preservation (ICBP),
(Now Birdlife International)
Wellbrook Court,
Girton Rd.,
Cambridge CB3 0NA, UK

International Institute for Environment and Development (IIED),
3 Endsleigh St,
London WC1H 0DD,
UK

International Society for Ecological Economics,
c/o Professor R. Costanza,
Coastal and Environmental Policy Program,
Centre for Environmental and Estuarine Studies,
University of Maryland,
Box 38, Solomons MD 20688–0038,
USA

IUCN, see World Conservation Union

IUCN-CESP Working Group on Environmental Assessment and Resource
Economics,
c/o Dr Rudolf de Groot,
Coordinator Climate Change Research Centre,
Nature Conservation Department,
Agricultural University,
Ritzema Bosweg 32a,
6703 AZ Wageningen,
The Netherlands

International Waterfowl & Wetlands Research Bureau (IWRB),
Slimbridge,
Glos. G12 7BX,
UK

Joint Nature Conservation Committee,
Monkstone House,
City Rd,
Peterborough PE1 1JY,
UK

National Science Foundation (NSF),
Washington DC,
USA 20550

The National Wildlife Federation,
1400 16th St, NW,
Washington DC,
USA

Natural Environment Research Council (NERC),
Polaris House,
North Star Avenue,
Swindon,
Wiltshire SN2 1EU,
UK

Nature Conservancy Council
see English Nature or Joint Nature Conservation Committee

NOAA/National Marine Fisheries Service,
Marine Entanglement Research Program,
7600 Sand Point Way, NE,
BIN C15700,
Seattle, Washington, 98115,
USA

Ramsar Convention Bureau,
Avenue du Mont-Blanc,
CH–1196,
Gland,
Switzerland

Research Institute for Nature Management (Netherlands),
Postbus 46,
3956 2R Heersum,
The Netherlands

Royal Society for the Protection of Birds (RSPB),
The Lodge,
Sandy,
Beds SG19 2DL,
UK

Smithsonian Institution,
1000 Jefferson Drive,
SW, Washington, DC, 20520,
USA

Tree Council,
35 Belgrave Square,
London SW1X 8QN.
UK

UNESCO,
UNESCO House,
7 Place de Fontenoy 75700,
Paris,
France

United Nations Convention on Environment and Development (UNCED),
Secretariat,
160 Route de Florissant,
Case Postal 80,
CH 1231,
Conches,
Switzerland

United Nations Environment Programme (UNEP),
PO Box 30552,
Nairobi,
Kenya

US Environment Protection Agency,
401 M Street, SW,
Washington, DC 20460,
USA

US Fish and Wildlife Service,
(Dept. of the Interior),
Washington DC, USA

The World Bank,
(Office of Environmental and Scientific Affairs),
1818 H St, NW,
Washington DC 20433,
USA

World Commission on Environment and Development (WCED),
Palais Wilson,
52 Rue des Pagnis,
Ch–1001,
Geneva,
Switzerland

World Conservation Monitoring Centre,
219 Huntingdon Rd,
Cambridge CB3 0DL,
UK

World Conservation Union (IUCN)
Rue Mauverney 28
CH–1196 Gland,
Switzerland

World Heritage Convention,
The Secretariat of the World Heritage Committee,
UNESCO,
7 Place de Fontency,
75700 Paris,
France

World Resources Institute (WRI),
1709 New York Avenue, NW,
Washington, DC 20006,
USA

Worldwatch Institute (WWI),
1776 Massachusetts Ave, NW,
Washington, DC 20036,
USA

World Wide Fund for Nature (WWF)
1. UK
World Wide Fund for Nature,
Panda House,
Ockford Rd,
Godalming,
Surrey GU7 1QU,
UK

2. USA
WWF
1250 24th St,
NW Washington, DC 20037,
USA

3. International
WWF,
Avenue du Mont Blanc,
CH–1196,
Gland,
Switzerland

Appendix C: Glossary

Adaptation The way in which an organism has evolved to become fitted for its way of life in terms of its behaviour, morphology, physiology etc.

Adaptive radiation Evolutionary diversification of a taxonomic group into different forms.

Agricultural ecosystem (agro-ecosystem) A human-made ecosystem developed and controlled by humans for production of agricultural crops.

Allele Different forms of a gene occupying the same locus on homologous chromosomes which undergo meiotic pairing.

Arthropod An animal with an external skeleton and jointed legs: insects, crustaceans, millipedes, centipedes etc.

Autotroph (autotrophic) An organism capable of synthesis of organic compounds from inorganic molecules using energy derived from chemical energy of inorganic compounds (green plants, sulphur bacteria).

Biodatabase An organized store of biological information in a form suitable for storing and analysing in a computer.

Biodiversity (Biological diversity) The degree of nature's variety, including both the number and frequency of ecosystems, species and genes in a given assemblage.

Biogeography The study of the geographical distribution of organisms and their habitats.

Biological conservation Biological conservation is an activity which aims to conserve living material including genetic material, populations and communities. This aim is achieved with the application of various sciences including conservation biology.

Biomass The weight of living material usually expressed as weight of dry matter per unit area (for example, g/m^2).

Biome A major region of the world defined by the climax vegetation and climate, for example, Taiga biome is the area of northern coniferous forests; temperate biome is the area with temperate climate and where there are or were mixed deciduous forests.

Biosphere That part of the earth and the atmosphere which is able to support life; the global ecosystem.

Biota A general term for all living organisms.

Biotope A term used by some landscape ecologists, meaning the smallest unit of a landscape that can be identified by the nature of the biota, especially the floristic characteristics (see ecotope).

Buffer zone In ecology a buffer zone is an area or zone which protects a habitat from damage, disturbance or pollution. It is an area (human-made or natural) which is managed so as to protect the ecological 'integrity' of an area.

Calcicolous (of plants) Tolerant of high levels of chalk or lime (cf. Calcifuge, acid-tolerant plants).

Carrying capacity The maximum number of organisms or biomass which can be supported in a given area under defined circumstances.

Chaparral Evergreen sclerophyllous vegetation found in southern California, which has adapted to dry conditions.

Cladistics, cladistic method This is an evolutionary classification method which is based on phylogenetic hypotheses and recent common ancestries rather than on phenetic similarity (phenetic – similarity based on characters selected without regard to evolutionary history and including characters arising from common ancestry).

Climax community The final stable community that is the termination of ecological succession.

Community Populations of different species inhabiting the same area or habitat, bound together by their biotic interrelationships.

Conservation biology Applied aspects of biology (taxonomy, genetics, ecology, biogeography) dealing with the conservation and management of living organisms, habitats, communities and ecosystems.

Conservation gain Return of a developed area to wildlife, change of land use to conservation, or creation of an artificial habitat.

Coppicing An ancient craft and form of management in which certain kinds of tree species such as sweet chestnut (*Castanea sativa*), hornbeam (*Carpinus betulus*), and hazel (*Corylus avellana*) are cut at ground level on a cycle of about 7–10 years. Growth from the cut base (stool) is harvested as a crop.

Creative conservation The construction of new habitats and the introduction of plants and animals to the new habitat.

Cultivar A plant variety maintained by cultivation.

Debt swaps The means by which the external debts of a country are purchased in return for the profits being used for conservation purposes such as acquisition of land for a nature reserve.

Demography The study of populations by way of statistical analysis of birth rates, death rates, age structure and population movement (immigration, emigration).

Diversity See Biodiversity.

DNA fingerprinting DNA (deoxyribonucleic acid) is the main constituent of chromosomes of all organisms which contain genetic codes unique to each individual organism. A unique 'fingerprint' can be obtained by extracting DNA from an individual, cutting the DNA using restriction enzymes, separating the fragments according to size by electrophoresis and detecting the pattern of sequence lengths.

The individual-specific sequence of DNA fragments of varying size is known as a 'DNA fingerprint'.

Ecological evaluation This is undertaken to identify the conservation needs of a species or the conservation importance of an area.

Ecological assessment For a species, this is undertaken to provide ecological data which can be used in the management and conservation of that species. For a community, this is undertaken to provide data which may help to identify the likely effects of pollution and other impacts.

Ecosystem Communities of organisms interacting with the abiotic environment as an ecological unit.

Ecology The scientific study of the distribution of organisms in time and in space and the study of interactions between organisms and their environment.

Ecotope A habitat type within a larger geographical area. The ecotope-system is one in which small-scale ecosystems or ecotopes are classified using biotic and abiotic factors that determine the species composition of the ecosystem.

Eigenvalue This is the proportion of the total variance which is accounted for by the corresponding principal component (in principal component analysis).

Endemic An organism which is found only in the area being considered (see Indigenous).

Environment All the components of an ecosystem that interact with living organisms: it includes the biotic (living) component and the abiotic (physical and chemical components).

Ethnobotany The study of man's cultural awareness of the values and uses of plants.

Extinction A species becomes extinct when there are no longer any living representatives.

Exotic species A species introduced from another region.

Eutrophication The nutrient enrichment of bodies of water caused by organic enrichment. Although a natural process, rapid eutrophication can drain oxygen levels and thus result in mortality of aquatic organisms.

Fauna A collective term for animals.

Faunal collapse Animal species extinctions or species loss in a particular area such as in a nature reserve.

Fen A mire (ecosystem with vegetation in wet peat) which is eutrophic.

Flora A collective term for plants.

Gene This is the basic unit of inheritance composed of specific sequences of material on a DNA chain, that has a specific locus on a chromosome.

Gene pool The total of all the genetic material in a breeding population.

Genetics The science of variation and heredity.

Genetic drift The occurrence of random changes in the genetic frrequencies of small isolated populations which are not due to selection and mutation.

Genetic engineering The technology of altering genetic material by artificial means.

Genetic fingerprinting See DNA fingerprinting.

Guild A group of species with similar ecological requirements and similar foraging strategies.

Habitat The locality or area used by a population of organisms and the place where they live.

Houb A damp area (usually freshwater) which is contained or enclosed behind a gravel spit.

Hydrology Science of the properties of water especially with regard to movement of water on land.

Indigenous Native to a particular region.

Management Management in the sense of management for conservation refers to activities undertaken to change or alter populations and communities of plants and animals. For example, mowing or grazing a grass sward could be a form of management adopted to change the species composition and increase the plant species richness. Another example could be the culling of certain age-classes of a mammal to ensure that populations are supported by the available resources such as food, water and breeding sites.

Metabolic pathway The sequential series of reactions by which one organic compound is converted into another in cells. Pathways may result in the formation of complex molecules and consume energy (anabolic pathways) or may involve the breakdown of molecules with release of energy (catabolic pathways).

Mire Collective term for bogs and fens; ecosystems where plants are rooted in wet peat.

Monotypic A taxonomic group with a single component such as a family or genus composed of a single genus or species respectively.

Native (plant or animal) A species which is characteristically found in that area and was not introduced.

Natural resource Naturally occurring resources include biological resources (living organisms) and non-biological resources such as water and minerals.

Natural area (naturalness) This refers to those areas not changed or affected by humankind. In effect, there are few such areas but there are many areas which could be called semi-natural, that is 'slightly' modified or affected by humankind.

Nature conservation The conservation of nature (wild organisms).

Niche The 'space' occupied by and the resources used by a species. Conceptually the niche has many dimensions and each resource used by the species can be considered as a dimension.

Pathogenic Capable of producing a disease.

Pesticide A chemical used to kill pests.

Phytosociology (plant sociology) The study of plant communities based on their classification, interdependence and association.

Plagioclimax A vegetation community which has formed as a result of certain landuse practices or traditional management. For example some grasslands are managed (mown, grazed or burnt) in such a way that further succession is prevented.

Population A collection of individuals (plants or animals) of all the same species living in a prescribed area.

Primary productivity The rate at which energy from the sun is absorbed by plants in the production of organic matter.

Race A specific subset of organisms within a species.

Radionuclide (radioactive isotopes) Isotopes are nuclides of an element having the same atomic number (or number of protons and hence chemical properties), but different mass numbers (total number of protons plus neutrons in the nucleus). They are specified by chemical symbols and mass numbers, hence ^1H, ^2H, are isotopes or radionuclides of hydrogen. Radionuclides emit radioactivity.

Reproductive potential The number of offspring produced by one female.

Seral stage The developmental stage of ecological succession.

Species A group of organisms of the same kind that can reproduce sexually among themselves but are reproductively isolated from other organisms. The basic unit in a classification. For example the scientific name for the African violet is *Saintpaulia ioantha*: *ioantha* is one species in the genus *Saintpaulia*.

Species composition The list species or assemblage of species present.

Species diversity. The variety and relative number of species in an area (c/f species richness).

Species richness The number of species present in an area.

Structure (woodland or vegetation structure) Distribution of vegetation biomass in the community. Often in distinct layers or strata. The complexity of the structure may influence the number of animal species in the vegetation.

Subspecies These are populations of the same species but with different gene pools and which may be geographically isolated. Subspecies can potentially interbreed and may do so where the population range overlaps.

Succession (ecological succession) The process or sequence whereby one type of community replaces another and eventually leads to the climax community (e.g. grassland leading to scrubland, open woodland and then closed forest).

Sustainable development One of many definitions is: development which meets the needs of the present without compromising the needs of the future generations to meet their own needs.

Symbiosis Two interacting organisms (symbionts) which live together and which benefit from each other are said to be living in a state of symbiosis.

Systematics The classification of organisms based on variation.

Taxonomy The scientific study of the description and variation of both living and extinct organisms.

Tetrad A square 2 km by 2 km used in mapping the distribution of a species.

Tombolos A sand or gravel spit which is similar to the shore and which may include a small island.

Watershed The catchment area for rivers and streams below the source streams.

Wetland An ecosystem where the substratum is permanently waterlogged. The Ramsar definition is as follows: areas of marsh, fen, peatlands or water, whether natural or artificial, permanent or temporary, with water that is static or flowing, fresh, brackish or salt, including areas of marine water the depth of which at low tide does not exceed six metres.

Woodland structure See structure.

Index